# Social Data Analytics in the Cloud with AI

The rise of cloud computing and Generative artificial intelligence (AI) has revolutionized data analytics pipelines. Analysts can collect, store, and process vast datasets in the cloud with high availability and scalability, and also leverage Generative AI to query and visualize datasets in natural languages. This pioneering textbook provides a gateway for students, educators, and professionals to develop and enhance social data analytics capabilities with the latest cloud computing and AI technologies. The textbook introduces educational cloud resources from leading technology companies, begins with foundational concepts, and progresses to advanced techniques.

**Features**

- The first textbook on cloud-based social data analytics with the assistance of Generative AI.
- Introduces educational cloud resources from leading technology companies like AWS, GitHub, and MongoDB.
- Presents a fully AI-powered data analytics pipeline from Python coding to data collection with APIs, cloud-based data storage, natural language queries, and interactive visualization.
- Analyzes Census and social media data with the latest large language models (LLMs).
- Provides hands-on exercises with real-world datasets on timely issues.

This textbook is an excellent resource for upper-level undergraduate and graduate students taking GIS, Urban Informatics, Social Science Data Analysis, and Data Science courses; faculty members teaching such courses; and professionals and researchers interested in leveraging cloud computing and Generative AI in social data analytics.

# Social Data Analytics in the Cloud with AI

Xuebin Wei and Xinyue Ye

CRC Press
Taylor & Francis Group
Boca Raton London New York

CRC Press is an imprint of the
Taylor & Francis Group, an **informa** business

Designed cover image: © Shutterstock, Khakimullin Aleksandr

First edition published 2025
by CRC Press
2385 NW Executive Center Drive, Suite 320, Boca Raton FL 33431

and by CRC Press
4 Park Square, Milton Park, Abingdon, Oxon, OX14 4RN

*CRC Press is an imprint of Taylor & Francis Group, LLC*



ISBN: 978-1-032-56905-5 (hbk)
ISBN: 978-1-032-56906-2 (pbk)
ISBN: 978-1-003-43761-1 (ebk)

DOI: 10.1201/9781003437611

Typeset in Palatino
by Deanta Global Publishing Services, Chennai, India

Access the Support Material at
https://www.routledge.com/9781032569055
All the Python codes are available at
https://github.com/xbwei/Social-Data-Analytics-in-the-Cloud-with-AI
Video tutorials and lectures on cloud-based data analytics with AI are available at
https://www.lbsocial.net

# Contents

# *Foreword*

**By Michael Batty**

Contemporary society across the entire globe is now dominated by ever-faster revolutions in information technology that involve the way computation is spreading out into every facet of urban life. Cities are becoming smart with our own behaviors being informed continuously by information being transmitted to personal devices that are always switched on. Vast data resources are now available remotely in what is now called "The Cloud," which is an amorphous constellation of data banks and archives that are continually being replenished by data from sensors embedded into every facet and fabric of our society. Access to this data is by no means uniform and different expertise is being deployed to use it in many different contexts, but layered on top of all of this is new software that in itself displays a degree of intelligence that enables us to make better sense of the world around us. This kind of artificial intelligence (AI) is the logical outcome of 75 years of software engineering that began when computers were first developed. Although AI is currently 'weak' in its embrace of 'human intelligence' – it is more appropriate to call it 'machine intelligence' – there is a continued transformation of our world using such software and its data that is enriching our behaviors and making our societies more prosperous, notwithstanding the darker side of such technologies that continue to invade our privacy in diverse ways.

When digital computing began in the middle of the last century with their simultaneous invention in the US, Britain, Germany (and some also say the Soviet Union) (Batty 2024), there had been glimpses of organizing calculations in algorithmic form since the Renaissance in Europe and probably as far back as the Greeks. Indeed, many ancient civilizations invented quite sophisticated number systems that assumed an order that could be manipulated using functions that are now intrinsic to our contemporary digital world. These functions are referred to as algorithms – procedures which do not necessarily imply numerical processing but are sets of instructions that deal with logical orders, applicable to many problems that are as qualitative as they are quantitative. In short, the invention of the digital computer heralded in the idea of the universal machine. Alan Turing, amongst others such as Kurt Gödel and John von Neumann, demonstrated that digital computation followed a logic where any problem or artefact could be reduced to the logic of the binary code; and in this way, they argued that computers encapsulated a most basic representation that could be extended to a multitude of problems and systems. Most of us, even those close to the invention of the digital computer, failed to appreciate this, and it took many years before computers were developed to embrace this greater multitude. Vision, sound, and intelligence itself (which we now call "weak" AI) have thence come within the ambit of digital computation.

During the last 75 years, computation has changed from a concern for hardware to one of software, and thence to data with AI now forming the cutting edge of a concern for how we might embed computers into every facet of our lives. But it is only since computers have scaled to the point where they can be used continuously and have become embedded into many routine functions that define the fabric of our society, that this area has really begun to explode. This is due to the confluence of two converging trends. First, as computers have scaled down to the smallest of devices – those that are handheld and those whose chips can be embedded into much smaller objects than the person, new streams of data about social and political systems as well as the economy, are being generated. These streams occur continuously in real time and at fine human and spatial scales, and they produce what has been called 'big data,' although we can only determine 'how big' once the sensors generating them have been switched off. The second trend is the speed at which contemporary computers function. Moore's Law (Moore 1965) suggests that the speed and memory of the most basic microprocessor is still doubling every 18 months. This is transforming everything that uses digital computation, and it has made possible digital applications that half a century ago were unimagined – in art, literature, music, and medicine as well as in the most routine data functions that pertain to everyday life such as in education, retailing, and all forms of communications and transport. The universal machine has thus become a reality.

You can see this best in contemporary software which the authors of this book introduce to those of us who are attempting to understand and model the social world using new methods of analysis and synthesis that explain our changing behaviours, particularly in space. As software has built up, layer upon layer, it has become ever more user friendly. Take, for example, geographic information systems (GIS). Most model software now can be generated as individual modules that are constructed in such a way that you can put what you need together quickly and almost effortlessly using languages such as R and Python. Graphics are easy to generate on the fly, in Jupyter notebooks for example; while access to enormous databases as Open AI systems (such as ChatGPT) are beginning to inform all kinds of analytical and modeling tasks.

The merger of networked computing and software development that are sustained by extensions of computation to all kinds of social entities, human physiologies, and political organizations, is generating societies that are fundamentally different from anything that has happened before. This is what is increasingly call the 'digital transformation' that is affecting everything we touch but layered on top of basic human functions, not replacing them but augmenting them. Computational social analytics has taken over 75 years to emerge and only now is there a real sense in which this is taking off. This book is testament to how we are able to blend and merge these changes together. In fact, programming for social analytics is a movement that is still in its infancy but the contributions here are simply the tip of the iceberg that is growing rapidly as new data sources emerge, new web resources become available, and as new models and theories based on new developments in both theory and data science rapidly evolve. This can be seen in what Klaus Schwab (2017) has called the *Fourth Industrial Revolution*, the first being mechanical, the second electrical, the third computational, and the fourth biological with a series of revolutions continuing and collapsing into one-another, heralding a transformational era where all organizations in society are becoming digital. As these successive revolutions are getting shorter and shorter and faster and faster, they are merging and converging with one another in such a way that social life is becoming unstable and liquid in a way that both fragments as well as integrates successive digital technologies. Social data analytics is just one corner of this new world, but many of its more generic features are evident in software systems that are described in this book.

There is already a strong bias in social data analytics to representing social systems in geographical terms. In fact, mainstream social science is only beginning to reflect this bias for geographical and more generally spatial representations of social phenomena have tended to remain somewhat detached from the key focus of social science. Political theory, sociology, and particularly economics have been quite resistant to the "spatial turn" in the social sciences, despite there being important subfields of urban sociology, politics, and economics, with relationships to different spatial scales from the local to the global. This resistance to merging the spatial with the social, political, and economic has been partly due to the fact that the geographical frame represents a stranglehold on ways of thinking spatially about systems. It has been hard to merge the idea of agency in social science with spatial aggregates that continue to dominate the geospatial world, although the development of new directions in behavioural economics and in agent-based modelling is beginning to provide much richer ways of thinking about social systems.

All these perspectives on how we can conceive the social world in software are to be found in the chapters of this book, but no book on this emerging subject area would be complete without some focus on the kinds of computation that characterizes this kind of social science. Since the development of widespread computation, the new foci of the field now revolve around major developments in statistical modelling, in linear and non-linear simulation that embrace new methods of estimating or training such models, in ways that enable them to learn. Machine learning at various depths has been stimulated by access to big data sets and there is a new-found focus on simulation that privileges big data and individual behaviours that enrich our traditional modes of aggregate modelling. New ways of dealing with quantitative data using natural language processing are becoming fashionable and giving us new insights into how textual bodies of data are generating new theories of human behaviour. The advantage of large language models – extensions of massive search of textual and visual as well as numerical data – are providing new insights into how social systems function, and although it is early days yet in this field, the terrain on which we are attempting to model social systems is changing fast. Weak forms of artificial intelligence are beginning to be used to enrich and augment both old and new models of how social systems function as well as enabling qualitative concerns to be more formally addressed.

In fact, the whole basis for our ability to predict or not in terms of social systems is under severe scrutiny as we begin to generate much deeper ways of learning how such systems function. Getting to grips with this complexity is essential for those involved with social data and its simulation, and this book represents a gentile and intelligible way in which we can learn about the opportunities and the limitations of new forms of computation from 'The Cloud' to 'AI' which will enrich our understanding of the social world.

## References

Batty, M. 2024. *The Computable City, Histories, Technologies, Stories, Predictions.* Cambridge, MA: The MIT Press.

Moore, G. E. 1965. Cramming More Components onto Integrated Circuit. *Electronics,* April 19, 114–117.

Schwab, K. 2017. *The Fourth Industrial Revolution.* London: Penguin Portfolio.

# *Foreword*

**By Budhu Bhaduri**

The field of data science and analytics has evolved tremendously over the past decade, driven by the exponential growth in data generation and advancements in computing technologies, including cloud computing. This transformation has enabled unprecedented opportunities for extracting meaningful insights from large datasets, revolutionizing and influencing decision-making processes. The book *Social Data Analytics in the Cloud with AI* by Xuebin Wei and Xinyue Ye stands as a testimony to these advancements, providing a comprehensive guide for leveraging the power of cloud-based data analytics. The book covers essential topics such as setting up cloud-based databases, data collection via APIs, and the use of Generative AI to enhance the data analytics process.

Data collection is a critical aspect covered in the book. The authors discuss three primary methods: downloading data directly from providers, web crawling, and using APIs. They highlight the advantages of using APIs, which offer flexibility in collecting specific data formats and integrating data collection into analytical pipelines for real-time analysis. The process of obtaining API tokens, querying data, and storing it in cloud databases is explained with practical examples. What makes this book extremely useful is its practical orientation. By providing access to demo code on GitHub and walking readers through real-world examples, Wei and Ye bridge the gap between theory and practice. This hands-on approach not only reinforces the concepts discussed but also empowers readers to apply their knowledge in practical scenarios, thereby fostering a deeper understanding of the material.

Generative AI, powered by foundation models, is presented as a transformative tool in the analytics workflow. The authors illustrate how AI can expedite data collection, cleaning, and visualization tasks, making the entire process more efficient. They also address the importance of ethical considerations in data analytics, emphasizing the need for responsible data usage and security measures to protect against cyber threats.

Moreover, the authors' commitment to keeping the content current is commendable. In a field that is rapidly evolving, the provision of ongoing updates and resources via their website ensures that readers have access to the latest techniques and tools in data analytics. This forward-looking perspective is crucial for staying relevant and competitive in the ever-changing landscape of data science.

In conclusion, *Social Data Analytics in the Cloud with AI* is an essential resource for students, researchers, and practitioners looking to harness the power of data in today's digital age. Xuebin Wei and Xinyue Ye have crafted a work that not only educates but also inspires, setting a high standard for future publications in the field. I am confident that this book will serve as a cornerstone for many successful data analytics endeavors and will be a vital addition to your professional library.

# *Foreword*

**By Michael F. Goodchild**

In the early 2000s the term "exaflood" became a popular way of describing the massive numbers of bytes of data that were being acquired, accessed, and shared through the Internet, the prefix "exa" denoting 1018. The flood has continued to grow exponentially since then, necessitating "zetta" (1021) and "yotta" (1024) and leading some to suggest the facetious "lotta." These data volumes present a very different world from that of the early days of computing, when scarce storage and computational power imposed strict limitations on what could be analyzed and for what purposes.

Yet analysis, or analytics, are how we derive value from data, and justify the enormous investments being made in imaging satellites, sensor networks, data centers, Internet connectivity, and computing devices – indeed, the entire computing infrastructure that is an essential part of our modern world. Data alone have little, if any, value, until they are processed, analyzed, visualized, and massaged into forms that can support decisions. Hence the often-cited definition of information, as data that are fit for purpose.

Some years ago the geospatial company Esri built its annual user conference around the mantra "Seeing what others can't," implying that only with the tools, infrastructure, and analytics of Esri's software could users gain the important insights that were hidden in the data. These are the insights that flow from the use of analytics and are the central focus of this book. Analytics take many forms and serve many purposes: they support the search for patterns that persist and repeat themselves in data, they help the analyst to search for causes that might explain and provide understanding, and they provide a basis for prediction based on past and current patterns. The methods of analytics have grown over decades and are encountered and learned by students at all levels, from the youngest to the most advanced.

The nature and goals of science have been evolving steadily and so have methods of analysis. The glory days of physics, with a steady stream of new discoveries of simple theories about the natural world, have been largely replaced by teams of investigators ransacking data for well-hidden and weak patterns, and nowhere is this more true than in the social sciences. Simple techniques such as regression have given way to advanced forms of machine learning, which are capable of finding weak relationships among large numbers of variables. These techniques must, out of necessity, be automated because of the sheer volume of effort that is required, and thus fall within the broad rubric of artificial intelligence (AI).

This notion of automating the research process has a long history, dating from the previous century, but has become more and more essential as data volumes increase and as the complexity of problems and issues multiply. It comes with its own limitations, however, and challenges many of the long-held assumptions of the scientific method. The findings of machine learning may not be readily understandable, though major efforts are underway in the research community to improve their explainability. As with even the oldest methods of analysis, the existence of a relationship between two variables does not necessarily indicate that one causes the other. And there is no guarantee that the results of analysis are replicable in other areas or at other times, casting doubt on the use of these methods for prediction. Uncertainty will often be present in the data used for machine learning, but suitable methods to incorporate it, report it, and visualize it are still few and far between.

In short, the methods of AI described in this book represent a powerful and exciting approach to social science in this era of abundant data and all-powerful computation, but must be approached with wisdom and an awareness of potential pitfalls. Generative forms of AI offer the long-sought potential for a plain-language interface to analytics, opening its use to researchers and students with limited knowledge of the internal workings of its methods, but amplifying the constant need for caution.

# *Preface*

In today's data-driven world, the ability to efficiently collect, analyze, and interpret vast amounts of social data is more critical than ever. However, analysts are always constrained by on-premises infrastructures to store and process big data or limited by computing skills to conduct sophisticated data queries or analyses. The rise of cloud computing and Generative artificial intelligence (AI) has revolutionized data analytics pipelines. Analysts can collect, store, and process vast datasets in the cloud with high availability and scalability and leverage Generative AI to query and visualize datasets in natural languages.

This textbook, *Social Data Analytics in the Cloud with AI*, provides a comprehensive introduction for students, educators, and professionals to develop and enhance social data analytics capabilities with the latest cloud computing and AI technologies. The textbook introduces educational cloud resources from leading technology companies, begins with foundational concepts, and progresses to advanced techniques. The content covers Python coding with AI-assistant, relational and NoSQL data storage, data collection with Application Programming Interfaces (APIs), data querying in Structure Query Language (SQL) and natural language, examining social media data with large language models (LLMs), and AI-powered data visualization with open resources and business intelligence (BI) tools.

Each chapter includes hands-on exercises with real-world datasets. The demo codes are published on the book's GitHub repository. Empowered by cloud computing and Generative AI, imagination is the only constraint in social data analytics!

# *Acknowledgments*

We express our heartfelt thanks to the anonymous reviewers for their invaluable feedback on our book. Their insights affirmed the timeliness and relevance of our work for both industry and academia. We are especially grateful for their suggestions to enhance real-world examples and address updates in online services, which have significantly enriched the quality of our book.

We also thank Dr. Michael Batty, Dr. Budhu Bhaduri, and Dr. Michael Goodchild for their thoughtful forewords. Their contributions highlight the critical role of Artificial Intelligence in understanding complex social systems, the rapid advancements in Cloud Computing and data science, and the evolving nature of data analytics. Their perspectives have significantly enriched the depth of our work.

Additionally, we would like to thank Ms. Irma Britton, Ms. Pradiksha Darshini, and Ms. Chelsea Reeves for their expertise and support throughout the publishing process.

Finally, we extend our gratitude to our family members, colleagues, and students for their encouragement and support, which have been invaluable throughout this journey. This work was partially supported by the Texas A&M Institute of Data Science Course Development Grant Program and the Harold L. Adams Interdisciplinary Professorship Research Fund.

# *About the Authors*

Dr Xuebin Wei is an associate professor at James Madison University (JMU) and an AWS Academy Certified Educator. He earned a Ph.D. in Geography from the University of Georgia in 2015. Dr Wei has developed a series of cloud-based technology courses that promote an engaged and inclusive learning environment and helps current students prepare for future careers. He was selected as a Faculty Ambassador of the AWS Educate Cloud Ambassador Program and received the JMU Excellence in Teaching & Learning with Technology Award in 2019.

Dr Xinyue Ye is the Harold L. Adams Endowed Professor in the Department of Landscape Architecture & Urban Planning, Department of Geography, and Department of Computer Science & Engineering at Texas A&M University–College Station, where he directs the Center for Geospatial Sciences, Applications, and Technology. Dr Ye received his Ph.D. in Geography from the University of California–Santa Barbara and San Diego State University in 2010. He is a Fellow of the American Association of Geographers and a Fellow of the Royal Geographical Society.

# 1

# *Introduction*

## Why Cloud Computing

Technology companies such as Amazon build and maintain vast computing infrastructures to support their daily business operations. These infrastructures are designed to handle peak loads, ensuring smooth and uninterrupted service during high-demand periods. However, during off-peak times, a significant portion of this computing power remains idle, resulting in underutilized resources. To maximize revenue and optimize resource utilization, these companies decouple the idle infrastructures from their primary applications and repurpose them as web services, collectively known as *cloud computing*. This strategic move allows them to offer scalable and flexible computing resources to external customers on-demand, transforming idle capacity into a valuable revenue stream while providing businesses and developers with the ability to scale their operations efficiently and cost-effectively. In 2006, Amazon Web Services (AWS)[1] started providing cloud computing services. As of March 2024, AWS offers over 200 services including computing, storage, networking, and analytics.[2] In addition, Google and Microsoft offer cloud computing services like Google Cloud[3] and Azure.[4]

Cloud computing services adopt a *pay-as-you-go pricing* model, where users are only billed for the seconds or minutes they use services (Amazon Web Services, Inc., n.d-a). This pricing model is a huge advantage of cloud computing services because users no longer need to overprovision computing infrastructure and can scale up or down the services based on real-time business demands. Users can also divide large computing tasks in parallel and utilize multiple cloud computing instances instead of one. Thus, they can complete the calculation tasks within a shorter time at the exact cost.

Cloud computing providers deploy their services across multiple regions and maintain the security and availability of their global infrastructures (Amazon Web Services, Inc., n.d-b). For example, AWS services are served

DOI: 10.1201/9781003437611-1

in over 240 countries and territories.[5] Each region has multiple Availability Zones with one or more data centers. Users can set up online businesses in areas close to their clients to reduce latency. They can back up their services in various Availability Zones to remain available if one or multiple data centers fail.

Cloud computing services can be classified into three categories: infrastructure as a service (IaaS), platform as a service (PaaS), and software as a service (SaaS).

- IaaS offers basic computing infrastructures like virtual machines, networking access, and data storage. Users have the greatest flexibility in building their applications and are responsible for updating the operating system and optimizing their applications. One example of IaaS is the Amazon Elastic Compute Cloud or EC2.[6] EC2 instances are the basic building blocks of AWS, and users can choose the latest central processing unit (CPU) or graphic processing unit (GPU) with different memory or storage sizes to build applications or train machine learning models.
- PaaS is designed for developers to focus on their applications without managing the underlying infrastructures. For example, AWS Cloud9[7] is a cloud-based integrated development environment (IDE) where developers can design, test, and deploy applications with a browser. They can choose the programming languages and don't need to install software development kits (SDK). Developers can also collaborate online and access other AWS resources, such as EC2 instances or databases.
- SaaS provides a complete product online and is typically charged on a subscription. Amazon QuickSight,[8] a cloud-based business intelligence (BI) tool, is a SaaS. Data analysts can use QuickSight to visualize data from various data containers and build interactive dashboards to identify patterns or obtain insights.

Cloud computing offers a versatile, budget-friendly, and robust platform for social science research. It enhances collaborative research capabilities, provides access to advanced data analysis tools, ensures data security, and supports the dynamic nature of social science studies. Cloud computing's ability to scale resources up or down is a significant asset in social science research. This flexibility is invaluable for handling diverse datasets and varying research demands, making it ideal for studies that range from small-scale surveys to large-scale social media data analysis. The cloud's pay-as-you-go model aligns well with the budget constraints often faced in social science research. It eliminates the need for significant upfront investments in data processing infrastructure, allowing for more efficient allocation of research funds. The cloud's accessibility is particularly beneficial for social science research, often conducted by teams across different institutions or geographies. By hosting data and applications in the cloud, researchers can easily share resources and collaborate effectively, irrespective of physical location. Cloud services often include built-in advanced analytics and artificial intelligence (AI) capabilities. These tools can be instrumental for social scientists, enabling sophisticated data analysis techniques such as predictive modeling. By leveraging these advanced technologies, social scientists can uncover deeper insights from social data, identify patterns and trends that may not be apparent through traditional methods, and enhance their research's overall rigor and impact. Additionally, the scalability and flexibility of cloud services allow researchers to handle large datasets efficiently, facilitating more comprehensive analyses.

The vast storage capabilities of the cloud simplify the management of large and diverse datasets, a common scenario in social science research. The datasets can include structured data (for instance, survey responses) and unstructured data (such as textual data from social media platforms). Security and data protection are critical in social science research, especially when handling sensitive information. Cloud providers typically offer robust security features, including data encryption and access controls, ensuring that research data is handled securely and ethically compliant. The high availability of cloud platforms ensures that research data and applications are accessible even during hardware failures, which is crucial for ongoing research projects.

Additionally, cloud services offer disaster recovery options, safeguarding against data loss and ensuring the integrity of research data. The cloud supports the rapid deployment of analytical tools, allowing social scientists to quickly adapt to new research methodologies or respond to emerging social trends. This agility is key in a field where staying abreast of societal changes is essential.

## Why Artificial Intelligence

Artificial intelligence (AI) development has reached a tipping point, with breakthroughs in Generative AI (GenAI) technologies occurring daily. Figure 1.1 illustrates the relationship between AI, machine learning (ML), deep learning (DL), and foundation model (FM). Generally speaking, AI refers to technologies and studies that enable computers or machines to perceive their environment and act like humans. Many AI applications, such as facial recognition and autonomous vehicles, are powered by machine learning models.

Machine learning is a subset of AI that employs algorithms and statistical models to identify patterns of a given dataset and make predictions or

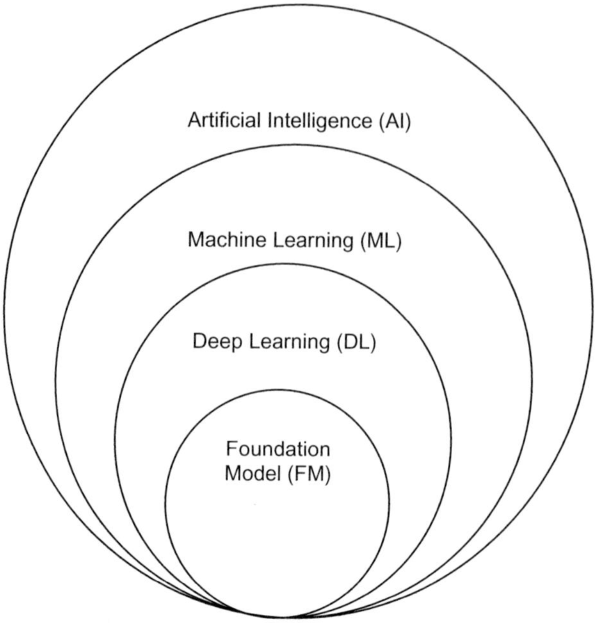


**FIGURE 1.1**
Foundation Models Are a Subset of AI

classifications on new datasets. There are three types of ML models: supervised, unsupervised, and reinforcement.

- Supervised learning: In supervised learning models, the correct answers, or labels, are provided in training datasets. If the labels are continuous numbers, the model is called *regression*. If the labels are categorical data, the model is called *classification*. The supervised learning models aim to find the best statistical relationships between the training datasets and the labels. When the model performance reaches an acceptable score, it can be deployed to make predictions on new datasets. Supervised learning models are the most popular type of ML model and are widely used in many industries and businesses, such as fraud detection, sales prediction, or image recognition.
- Unsupervised learning: The labels are not provided in unsupervised learning models. Instead, the models identify patterns in the training datasets and categorize similar data points into the same group, i.e., *clustering*. Unsupervised learning models can help identify data structures, recommend relevant products, or detect anomalies.
- Reinforcement learning: In reinforcement learning, an agent attempts to achieve a goal in a set environment through many

learning interactions. During the learning process, the agent is rewarded for good actions and punished for errors. Finally, the agent can correctly interact with the environment and find the best path to achieve the goal. The applications of reinforcement learning models include self-driving cars, robotics, and gaming.

Deep learning is a subset of ML that utilizes artificial neural networks (ANNs), which are more complicated but accurate in prediction and classification. An ANN mimics human neural networks and contains multiple layers. Each layer comprises several nodes, namely *neurons*, and is fully connected with neurons of other layers. Each neuron is assigned a weight called a *parameter*, and all the weights are adjusted during the model training process to increase the model performance gradually. Convolutional neural networks (CNN)[9] and recurrent neural networks (RNN)[10] are popular ANNs. A CNN contains a convolutional filter, which is helpful in image processing and recognition. In RNN, the output of a layer in the previous learning process becomes part of the layer's input in the following learning process. RNN is particularly useful in dealing with sequence data, such as texts.

Foundation models are the latest milestone of DL, and large language models (LLM) are one type of FM that focuses on natural language processing (NLP). FMs are highly complex and require massive training datasets. For example, GPT-2, a popular LLM, contains 1.5 billion parameters.[11] Traditional ML is task-specific, i.e., one model can only handle one type of job. FMs are normally pre-trained and can perform various tasks, such as text generation, sentiment analysis, language translation, etc (Amazon Web Services, Inc., n.d-c). Training a new FM can be time-consuming and resource-expensive. Therefore, organizations or companies that want to optimize the output of a pre-trained FM typically choose the retrieval-augmented generation (RAG)[12] technique to bring domain knowledge or internal datasets to the FM so that the model can produce more relevant content. The other methods to increase the model performance or efficiency involve prompt engineering and fine-tuning.[13]

Generative AI (GenAI) is powered by foundation models that can create new content, such as texts, images, or videos (OpenAI, n.d.). For example, ChatGPT is a popular chatbot developed by OpenAI.[14] People can communicate with ChatGPT in natural language, and ChatGPT can answer questions, draft emails, analyze data, and create charts. Amazon Q, an AI assistant developed by AWS,[15] can assist developers in writing or debugging codes, help data analysts create visualizations, or facilitate business owners to improve customer service.

With cloud infrastructures, everyone can access the latest Generative AI technologies with a browser. This book focuses on utilizing cloud computing resources and Generative AI technologies to analyze big social data, from data collection to query and visualization. AWS Academy,[16] DeepLearning.AI,[17] and LBSocial[18] provide excellent online courses for people interested in cloud computing or artificial intelligence. Grus's *Data Science from Scratch* is also an excellent textbook that teaches machine learning in Python (Grus 2019).

Analysts can focus on the problems they are interested in and leverage Generative AI to complete the associated computation tasks. Therefore, this textbook doesn't dive deep into technical details but introduces data analytics' fundamental components and principles, incorporated with the latest Generative AI applications in the cloud. The textbook also discusses the limitations of Generative AI in data analysis and how to conduct *responsible data analytics*. Advances in cloud technology and AI have streamlined the process, allowing individuals and organizations to leverage powerful analytical tools and vast computational resources without the need for extensive on-premise infrastructure. Analyzing data in the cloud with AI is becoming more accessible and exciting than ever before!

## Notes

1. https://aws.amazon.com/
2. https://docs.aws.amazon.com/whitepapers/latest/aws-overview/introduction.html
3. https://cloud.google.com
4. https://azure.microsoft.com/
5. https://aws.amazon.com/about-aws/global-infrastructure/
6. https://aws.amazon.com/ec2/
7. https://aws.amazon.com/cloud9/
8. https://aws.amazon.com/quicksight
9. https://developers.google.com/machine-learning/glossary/#convolutional_neural_network
10. https://developers.google.com/machine-learning/glossary#recurrent-neural-network
11. https://openai.com/index/gpt-2-1-5b-release/
12. https://aws.amazon.com/what-is/retrieval-augmented-generation/
13. https://platform.openai.com/docs/guides/optimizing-llm-accuracy/llm-optimization-context
14. https://openai.com/chatgpt/
15. https://aws.amazon.com/q/
16. https://aws.amazon.com/training/awsacademy/
17. https://www.deeplearning.ai/
18. https://www.lbsocial.net/

## References

Amazon Web Services, Inc. n.d.-a. "Generative AI on AWS – Generative AI, LLMs, and Foundation Models – AWS." Amazon Web Services, Inc. Accessed June 27, 2024. https://aws.amazon.com/ai/generative-ai/.

Amazon Web Services, Inc. n.d.-b. "Global Infrastructure - AWS." Accessed July 7, 2024. https://aws.amazon.com/about-aws/global-infrastructure/.

Amazon Web Services, Inc. n.d.-c. "Overview of Amazon Web Services - Overview of Amazon Web Services." Accessed July 7, 2024. https://docs.aws.amazon.com/whitepapers/latest/aws-overview/introduction.html.

Grus, Joel. 2019. *Data Science from Scratch: First Principles with Python*. 2nd edition. Sebastopol, CA: O'Reilly Media.

OpenAI. n.d. "Pioneering Research on the Path to AGI." Accessed June 27, 2024. https://openai.com/research/.

# 2

# *Set up a Free Cloud-based Learning Environment*

Cloud computing delivers various services, such as data analysis, storage, and networking, over the Internet. The leading providers in this domain are Amazon Web Services (AWS), Microsoft Azure, and Google Cloud Platform (GCP). Each of these giants extends complimentary credits to students and educators, promoting the academic application of their services. This provision enables students to harness cutting-edge technologies from their laptops, encompassing tools such as programming editors, big data storage, machine learning models, and even the latest large language models (LLM). For educators, cloud computing is not just a tool to diminish teaching expenses; it's a catalyst for fostering a more inclusive and diverse online learning landscape. This technological advancement facilitates an engaging and resource-rich educational environment accessible from anywhere.

This chapter guides educators through setting up a complimentary academic account on AWS and creating an online learning space via the AWS Academy program.[1] AWS Academy(AWS Academy, n.d.) will be the primary platform for Python programming, storing, processing, and analyzing data throughout this book.

In addition to AWS Academy, we will explore two prominent cloud-based platforms: MongoDB[2] and GitHub.[3] MongoDB offers a robust No-SQL data storage solution in the cloud that is adept at managing massive and diverse data. On the other hand, GitHub is a widely used cloud-based hosting service for Git repositories, serving as a popular hub where programmers share their projects and collaborate. Both MongoDB (MongoDB, Inc., n.d.) and GitHub (GitHub, Inc., n.d.) provide free resources to students and educators. In this chapter, we will delve into these complimentary offerings. Subsequent chapters will teach students how to use MongoDB to store and analyze social media data and leverage GitHub to share their work and build an impressive online portfolio.

## AWS Academy

AWS Academy is an Amazon Web Services program that intends to empower "higher education institutions to prepare students for industry-recognized


DOI: 10.1201/9781003437611-2

certifications and careers in the cloud."[4] This section introduces how institutions, educators, and students can join the AWS Academy program, benefit from the free teaching and computing resources, and receive badges and certifications from AWS Academy.

## Join the AWS Academy Program

To join the AWS Academy Program, an institution program administrator must complete an institution application, and each institution only needs to submit one application. Once the AWS Academy Program approves the application, all the students and educators of the institution can join the AWS Academy Program. The person applying by default will be to the Central Point of Contract (CPOC), responsible for communicating with AWS Academy and nominating educators. The application process involves submitting the application at the *AWS Academy website*,[5] attending a virtual Program Overview Session, reviewing and signing legal documents if approved, and obtaining access to the *AWS Academy Portal*.[6] Once an institution is enrolled in the AWS Academy program, the educators of the institution can be nominated by the CPOC. The nominated educators who acknowledge the Program Guide (AWS Academy 2019) and complete the required *Getting Started with AWS Academy Orientation* course will receive access to the AWS Academy portal and the *AWS Academy Learning Manage System (LMS)*.[7]

Both institution administrators and educators can utilize the AWS Academy Portal. Figure 2.1 displays the interface of the AWS Academy Portal. The *Home* page lists all the AWS resources educators can use in their classes, including the *ready-to-teach* courses and *AWS Learner Labs*.

- AWS develops and maintains those ready-to-teach courses, including lecture modules, hands-on labs, knowledge checks, and scenario-based projects. Currently, there are more than 15 ready-to-teach courses. The topics include Cloud Computing Foundations,

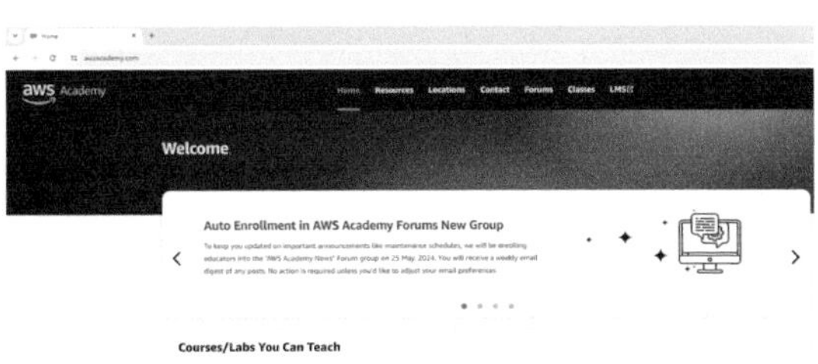


**FIGURE 2.1**
A Screen of the AWS Academy Portal

Cloud Architecture, Cloud Development, Cloud Operations, Data Engineering, Machine Learning, etc. The course contents are constantly updated, and new courses are frequently added. Educators can self-study those courses or deliver them directly in their classes. Both students and educators can access the designated AWS services in the hands-on labs with free AWS credits. Those cloud resources, such as Python editors or databases, will be terminated after the completion of each lab.

- Educators can create an Academy Learner Lab if they plan to design their labs using free AWS services. Academy Learner Labs are separate courses that provide access to over 100 AWS services. Students enrolled in a Learner Lab are allocated free credits, and the AWS resources will be preserved till the credits run out or the course ends. Many of the services are constrained. For example, students may only access standard or limited services on AWS and can not change the Billing management or Identity and Access Management (IAM). The available AWS services may change over time, and educators should check the Learner Lab's ReadMe file regularly.

The *Resources* page on the AWS Academy Portal provides detailed information about the ready-to-teach courses and learner labs, such as course outlines and readiness rubrics. These documents are also translated into multiple languages. Educators can find the CPOC contact information on the *Contact* page. To contact AWS Academy, educators need to submit an online ticket.[8] The AWS Academy *Forums* is an online community where educators and AWS Academy staff can ask and answer questions, announce new courses or features, and share teaching experiences or resources. The *Classes* page lists the educator's upcoming courses or labs, courses in delivery, and past courses. The *LMS* page can direct educators to the AWS Academy LMS, where educators and students can manage or access the AWS courses and labs. Table 2.1 summarizes the purposes and intended users of the AWS Academy Website, AWS Academy Portal, and AWS Academy LMS.

## Create AWS Academy Courses

The AWS Academy LMS website is built on Canvas.[9] Figure 2.2 shows the interface of the LMS. Like other Canvas LMS, educators can create and manage courses, invite students, grade assignments, and communicate with students. Students can access course content, complete labs and quizzes, and check their grades. However, even though both use a Canvas system, the AWS Academy LMS is separate from the institution's LMS. Therefore,

**TABLE 2.1**

Summary of the AWS Academy Websites

| Website | Purpose | Intended Users |
|---|---|---|
| AWS Academy Website | Fill out the institution's application | Administrators |
| AWS Academy Portal | Nominate educators;<br>Access AWS Academy resources and forums | Administrators and Educators |
| AWS Academy LMS | Create or access the ready-to-teach courses or Learner Labs | Educators and Students |

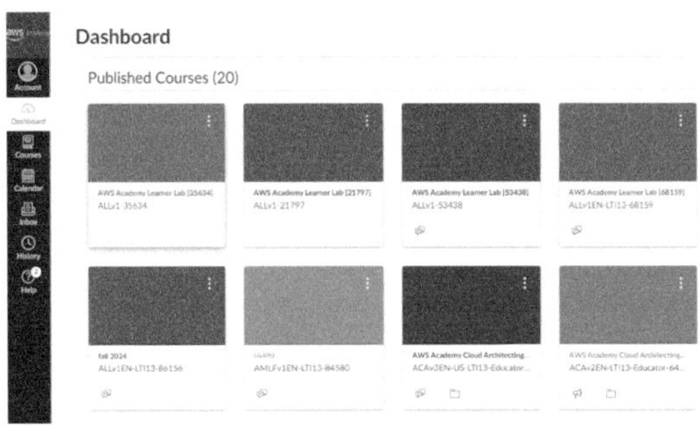


**FIGURE 2.2**
Interface of the AWS Academy LMS

educators can check students' progress and grades on the AWS Academy LMS but may need to copy students' grades to their institution's LMS manually if completing AWS lectures or labs is part of the course assignments.

To create a course or a Learner Lab, educators need to open the *Account* page and click the *Create a Class* link. On the *Create Course* page (Figure 2.3), educators can choose one of the ready-to-teach courses or the Learner Lab. They also need to specify the start date and time, the end date and time, time zone, language, and delivery modality, i.e., in-person, entirely online, or blended. Please note: As announced on July 1, 2024,[10] the maximum duration of any courses or labs is *six months,* and students receive *$50* credits in a Learner Lab. They will lose access to the lab after six months or spend all the credits.

The created course or lab will generally be available on the requested start date. Once the course is ready, educators can open the *People* page in the created course and invite students to join the course. Figure 2.4 shows the *Add People* window, where educators can invite their students by providing their email addresses. The invited students will receive an email from AWS Academy with an invitation link. Students can click the link, create their account on AWS Academy LMS, and access the course or labs created by their instructors. Students who use Canvas as their institution's LMS may be confused

**FIGURE 2.3**
Create AWS Course Window

**FIGURE 2.4**
Invite Students to Join an AWS Course

when creating an AWS LMS account. Educators may need to remind students that the AWS LMS differs from their institutions' LMS and that they can't access the AWS Canvas with the institutions' Canvas credentials.

## Earn AWS Learning Badges

Educators or students who complete an AWS Academy read-to-teach course can earn the corresponding AWS Learning Badge.[11] The AWS Learning Badges are free to claim and a great way to demonstrate knowledge or skills in cloud computing, data analytics, or machine learning. AWS badges are issued and hosted on Credly[12] and can be shared digitally, such as on social media or online profiles.

In addition, the AWS Academy courses are designed to match the structure of AWS Certification,[13] an industry-standard and well-recognized credential. Although earning a learning badge doesn't guarantee passing the corresponding certificate exam, the Academy courses help students prepare for the associated certificate exams (Amazon Web Services, Inc., n.d.). For example, students who complete the AWS Academy Cloud Foundations course can receive the Cloud Foundations learning badge, as shown in Figure 2.5. The

**FIGURE 2.5**
AWS Academy Cloud Foundations Badge

**FIGURE 2.6**
AWS Cloud Practitioner Certificate

students can use the resources provided in the course, including the exam guidelines, sample questions, and practice exams, to better understand the AWS Certified Cloud Practitioner exam. Students who pass the certification exam can earn the AWS Cloud Practitioner certificate, as shown in Figure 2.6. AWS learning badges don't expire. However, AWS certificates typically expire in 3 years. Please note: Students can take any AWS Academy courses and receive the associated badges once they have passed the online course. The AWS certificates require a deep understanding of the relevant AWS services and multiple years of working experience. People must earn the foundational certificates before taking the professional, associate, or specialty exams.

AWS Academy provides two special badges for educators. Educators who are enrolled in the Academy program can receive the AWS Academy Educator Badge, as shown in Figure 2.7. Educators who actively deliver AWS Academy courses and earn a valid AWS certificate can receive the AWS Academy Certified Educator Badge, as shown in Figure 2.8. Both educator badges expire in 2 years.

## MongoDB

Social data analytics might require substantial big data storage capabilities. This is because the analysis of social data involves processing vast

**FIGURE 2.7**
AWS Academy Educator Badge

**FIGURE 2.8**
AWS Academy Certified Educator Badge

amounts of information generated from diverse sources. Efficient storage solutions are essential to handle the volume, variety, and velocity of this data, enabling effective analysis and the extraction of valuable insights that can inform decision-making and strategy development. MongoDB[14] is ranked as number 5 among all databases and the *number 1 NoSQL database*.[15] Compared with relational databases that only store structured data, MongoDB can store flexible semi-structured data, such as JSON documents, and scale horizontally at lower costs (MongoDB, Inc., n.d.). Therefore, MongoDB is more suitable for quickly storing and processing diverse and large-volume data.

Registering a MongoDB account or deploying an M0 cluster is "free forever."[16] MongoDB also offers training resources for educators, students, and the public, such as MongoDB University,[17] Certification Program,[18] and Educator Program.[19]

- MongoDB University provides free learning courses and hands-on labs for all users. Users can customize their learning path based on expertise levels, programming languages, and MongoDB products. These courses can help users better prepare for the MongoDB Certification exams.
- Like AWS, MongoDB offers certifications for people who want to demonstrate their data analytical capabilities. By passing the certification exams, people can earn the corresponding badges, which are also listed on Credly.[20]

- The MongoDB Educator Program allows educators and students to apply for MongoDB credits, take free certification exams, and join global academic communities. Educators can also access curriculum resources crafted by MongoDB experts.

This book uses the MongoDB M0 cluster and recommends MongoDB Atlas,[21] the cloud-based data platform, to build and manage MongoDB databases. Educators can sign up for a free MongoDB account, log in to MongoDB Atlas, and create *Organizations* to manage students' projects in different courses. For example, educators can create an *organziation* for a data mining course, as shown in Figure 2.9.

Once an organization is created, the creator is the organization's owner and can invite students enrolled in the data mining course to the same organization. As shown in Figure 2.10, educators can open an organization's *Access Manager* and type students' email addresses. Educators should ensure that each student is granted permission only to be an *Organization Member* and *Organization Project Creator*. Those permissions allow students to create

**FIGURE 2.9**
Create a MongoDB Organization

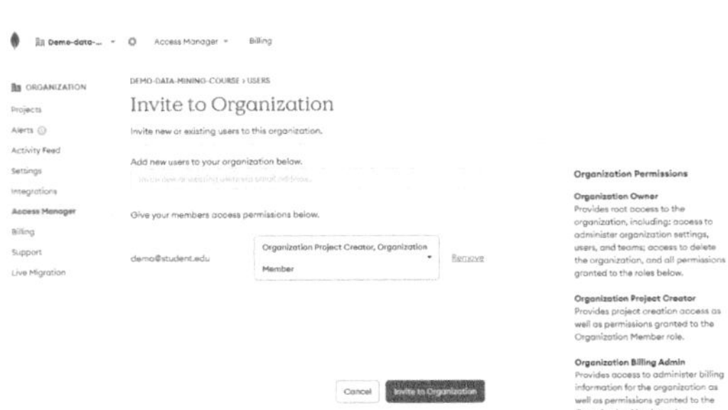


**FIGURE 2.10**
Invite Students to a MongoDB Organization

individual projects and databases within the organization but block access to other students' projects.

Invited students will receive an invitation email to register a MongoDB account and create projects in the invited organization. Creating organizations and projects is free, and organizations help educators efficiently manage students in different classes. As the owner of the organizations, educators can access each student's project and database to check their work and provide assistance when necessary. Students can manage their projects and databases with a browser but can't access other students' work.

## GitHub

GitHub[22] was originally a code hosting platform for developers to store, collaborate, and version control their projects. It is also the largest online community for developers to share their coding-related works, resources, datasets, etc. Students or educators can access the free resources on GitHub to advance their knowledge of data science and publish their work on GitHub to build online portfolios. For example, students taking a Python programming course can link their Python editor with a GitHub repository. Every time they complete a lecture or lab, they can synchronize their codes and datasets to the linked repository. Students who need to review their codes can check them on GitHub without running their Python editor. Students can back up their codes on GitHub when working on a capstone project. They can collaborate online and test different versions of the codes with the version control functions on GitHub. Finally, students can publish their repositories to demonstrate their coding or data analysis capacity.

GitHub has expanded its services from simple code hosting to developer platforms (GitHub, Inc., n.d.). For example, GitHub Codespaces[23] is a cloud-based development environment where users can write, execute, and debug codes in a browser, collaborate online, and manage product versions. GitHub Copilot[24] is an AI assistant that can complete codes automatically, generate new codes based on users' inputs, and review existing codes to provide recommendations or identify vulnerabilities.

Depending on the size of users, individuals or organizations can create *personal* accounts, *organization* accounts, or *enterprise* accounts on GitHub.[25] By default, each GitHub user owns a personal account that can manage individual resources, such as repositories or datasets, and collaborate with other users. GitHub organization accounts are shared accounts that allow multiple users to collaborate on various projects. Personal accounts in an organization can have different roles or access various organizational resources. Enterprise accounts enable administrators to manage multiple GitHub organizations, configure billing and usage, and access advanced security features.

GitHub offers free and paid plans for different types of accounts.[26] A personal account user can use the free plan called *GitHub Free,* which includes unlimited public or private repositories, 120 Codespaces core hours, and 15 GB of Codespaces storage per month. The paid personal account is named *GitHub Pro,* which has 180 Codespaces core hours, 20 GB of Codespaces storage per month, and advanced tools in private repositories, such as insights graphs and auto-linked references. Organization account users can also choose the GitHub Free or upgrade to paid plans, including *GitHub Team* and *GitHub Enterprise.*

Like AWS and MongoDB, GitHub has a dedicated program to support students and educators named GitHub Education.[27] Students who join the GitHub Education program can receive a free GitHub Pro account. They will also receive the Student Developer Pack[28] containing cloud computing resources and educational content that GitHub's industrial partners provided. Educators verified by the program can create GitHub Team accounts and use Copilot for free. They can also use GitHub Classroom[29] to create virtual classrooms, design coding assignments with template repositories, and automatically grade student submissions. The GitHub Education program also provides free GitHub Enterprise accounts and faculty training for institutions.

As the world's most comprehensive cloud provider,[30] AWS provides alternative solutions like MongoDB and GitHub. For example, DynamoDB[31] and DocumentDB[32] support NoSQL data storage. AWS CodeCommit[33] can host Git repositories. Amazon Q Developer[34] is also an AI assistant that can help with coding and security scanning. Educators are encouraged to select the resources that meet their needs and fit their expertise. This textbook mainly uses the available services in AWS Academy and chooses the MongoDB data storage and GitHub code repository services with free accounts.

## Notes

1. https://aws.amazon.com/training/awsacademy/
2. https://www.mongodb.com/
3. https://github.com/
4. https://aws.amazon.com/training/awsacademy/
5. https://www.awsacademy.com/InstitutionApplication
6. https://www.awsacademy.com/
7. https://awsacademy.instructure.com/
8. https://support.aws.amazon.com/#/contacts/aws-academy
9. https://www.instructure.com/canvas
10. https://www.awsacademy.com/forums/s/group/0F94N000000JPoVSAW/aws-academy-news

11. https://aws.amazon.com/training/badges/
12. https://www.credly.com/organizations/amazon-web-services/badges
13. https://aws.amazon.com/certification/
14. https://www.mongodb.com/
15. https://db-engines.com/en/ranking
16. https://www.mongodb.com/pricing
17. https://learn.mongodb.com/
18. https://learn.mongodb.com/pages/certification-program
19. https://www.mongodb.com/academia
20. https://www.credly.com/organizations/mongodb/badges
21. https://www.mongodb.com/atlas
22. https://github.com/
23. https://github.com/features/codespaces
24. https://github.com/features/copilot
25. https://docs.github.com/en/get-started/learning-about-github/types-of-github-accounts
26. https://docs.github.com/en/get-started/learning-about-github/githubs-plans
27. https://github.com/edu
28. https://github.com/edu/students
29. https://classroom.github.com/
30. https://aws.amazon.com/what-is-aws/
31. https://aws.amazon.com/dynamodb/
32. https://aws.amazon.com/documentdb/
33. https://aws.amazon.com/codecommit/
34. https://aws.amazon.com/q/developer/

## References

Amazon Web Services, Inc. n.d. "Earn and Claim Digital Badges | AWS Training | AWS." Amazon Web Services, Inc. Accessed June 28, 2024. https://aws.amazon.com/training/badges/.

AWS Academy. 2019. "AWS Academy Program Guide." https://uemgroup.s3.amazonaws.com/uploads/sites/3/2021/03/AWS-Academy-Program-Guide-20191203-v3.1-EN.pdf.

AWS Academy. n.d. "Get Started with AWS Academy." Amazon Web Services, Inc. Accessed September 30, 2024. https://aws.amazon.com/training/awsacademy/

GitHub, Inc. n.d.-a. "GitHub Education Documentation." GitHub Docs. Accessed June 28, 2024. https://docs.github.com/en/education.

GitHub, Inc. n.d.-b. "GitHub.Com Help Documentation." GitHub Docs. Accessed June 29, 2024. https://docs.github.com/en.

MongoDB, Inc. n.d.-a. "MongoDB Courses and Trainings | MongoDB University." Accessed June 28, 2024. https://learn.mongodb.com/.

MongoDB, Inc. n.d.-b. "MongoDB Documentation." Accessed June 28, 2024. https://www.mongodb.com/docs/.

# 3

# *Introduction to Python Programming and Data Analytics*

Data analytics involves fast data calculation and massive data storage. Therefore, basic programming techniques can largely benefit data analysis processes. This chapter teaches how to use Python, the number one programming language,[1] in data analytics. Python is an incredibly efficient language and easy to learn for beginners. The Python community contributes rich collections of libraries and resources that are widely used in big data analytics and machine learning models.

This chapter has two sections: The Python Tutorial section covers the basic programming techniques in Python such as data types, flow control, and program design. Users learn to employ AWS Cloud9,[2] a cloud-based integrated development environment (IDE), to write, execute, and debug Python codes and leverage the AI-powered coding assistant, Amazon Q Developer,[3] to create Python codes automatically. The Data Analytics section introduces the popular Python libraries in data science and demonstrates how to import, manage, and query data in JupyterLab,[4] a web-based interactive development environment.

Two textbooks are recommended for audiences wanting to dive deep into Python: *Think Python: How to Think Like a Computer Scientist* by Allen Downey (2015) and *Python Crash Course* by Eric Matthes (2019). The reference section also lists the official documents (Python Software Foundation, n.d.; pandas, n.d.; GitHub, Inc., n.d.) and tutorials (John Hunter et al., n.d.; Jupyter Team, n.d.-a; n.d.-b) for the tools introduced in this chapter.

## A Crush Python Tutorial

A program is a sequence of instructions that specifies machines to perform computations. It generally requires an *input*, conducts *calculations*, and produces an *output*. The input can be data typed via a keyboard, a file stored on a local hard disk, or datasets from a database in the cloud. The calculations involve mathematical operations such as addition or multiplication,

DOI: 10.1201/9781003437611-3

conditional executions, and repetitions. The output can be displayed on a computer screen, saved to a local file, or stored in a database.

Programming languages like Python are *formal languages* designed to express those computations. They have strict *syntax* rules to control the structure and format. Developers must understand and follow proper syntax in programming; otherwise, errors will occur. Adhering to the correct syntax ensures that the code is interpreted correctly by the compiler or interpreter, preventing runtime errors and bugs. By maintaining proper syntax, users can write clean, efficient, and error-free code, which is more reliable and maintainable. Users need to install an *interpreter* to execute the Python codes on computers.[5] The most prevalent version is Python 3, such as Python 3.12.4,[6] released on June 6, 2024. Code *editors* are generally recommended when writing Python or other codes, as they can recognize code structures, highlight sections, provide tips, and make codes easy to understand. They also offer *debugging* functions to help developers track down errors. Python's Integrated Development and Learning Environment,[7] or IDLE, is included in the Python interpreter. However, most programmers prefer advanced code editors like Cloud9 and Jupyter Notebook. This section introduces the basic syntax of Python with a Cloud9 editor and creates a GitHub repository to store and version control the Python codes. You are encouraged to follow the next section to create a GitHub repository and test the Python codes yourself. The textbook also hosts a repository[8] sharing all the demonstrated Python codes.

## Set up Cloud9 and GitHub Repository

AWS Cloud9 is a cloud-based integrated development environment (IDE) with a Python interpreter installed. In Cloud9, we can write, debug, and execute Python codes anywhere with a browser. As a fully managed cloud service, AWS maintains the instances that host the Cloud9 application and is responsible for upgrading the operating system and installing necessary packages. We can manage the instance remotely, such as installing packages or uploading datasets. No software needs to be installed on the local computers.

In addition, we can create a repository on GitHub. A *repository* is a fundamental element on GitHub that can store codes, datasets, and version history. We can roll back to a specific code version when needed, merge codes created by different users without conflict, and view the code and results without running a code editor. In Chapter 2, we made a free GitHub account and an AWS Academy Learner Lab. We can use the GitHub account to create a repository and set up the Cloud9 IDE in the AWS Learner Lab.

We need to log in to GitHub[9] and open the new repository page[10] to create a repository. As shown in Figure 3.1, we can provide a repository *name* and *description* and set the repository to be *public* or *private*. Since we want to use this repository as an online portfolio to demonstrate our data analytics capabilities, we have selected the public option. A *README* file is like the front page of a repository that can introduce the purpose of this repository, how the codes and datasets are organized, and a list of the contributors. The *.gitignore* file specifies what files will be ignored when we upload codes to our repository. For example, if we store database credentials in a local file, we can list that file name in the .gitignore, and GitHub will not upload that file to the repository. The .gitignore file is also language-specific, and we can choose a Python template for the .gitignore file. We should also select an appropriate *license* to protect our codes. The MIT license[11] can cover our needs in this scenario. However, users can check the *choosealicense.com*[12] to decide which license works best. Once the repository is created, we can upload Python codes, edit the README file, and track the changes in our repository.

To start the AWS Learner Lab, we need to log in to AWS Academy LMS[13] with a student account. Once logged in, we should see an *AWS Academy Learner Lab* course. Open the Learner Lab course, select *Modules*, and choose *Launch AWS Learner Lab*. We click *Start Lab* to access the AWS services. Figure 3.2 shows the interface when the lab is ready. There are $50 credits available in this lab. As we continue using AWS services, we will see the credits being consumed. To save the credits, we should *End Lab* promptly when we are not using AWS services. Please note: Some AWS services, like a database instance, will continue running and costing the credits even if we end the lab. We need to stop such services manually before we end the lab. The *Readme* file lists the available AWS services in this lab environment. It is critical to check the Readme file carefully and regularly to see what services are available and how to access them. AWS frequently changes the available services or service restrictions in the Learner Lab. The green dot next to the *AWS link* indicates the lab is ready. We can click the AWS link to open the AWS *Console* and access AWS cloud services, like Cloud9.

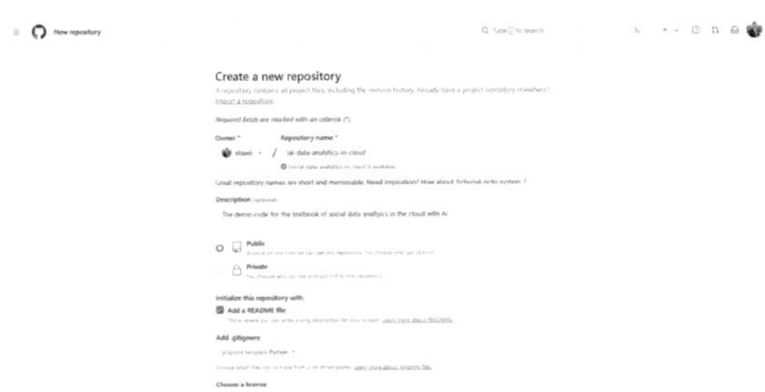


**FIGURE 3.1**
Create a GitHub Repository

**FIGURE 3.2**
Interface of AWS Academy Lab

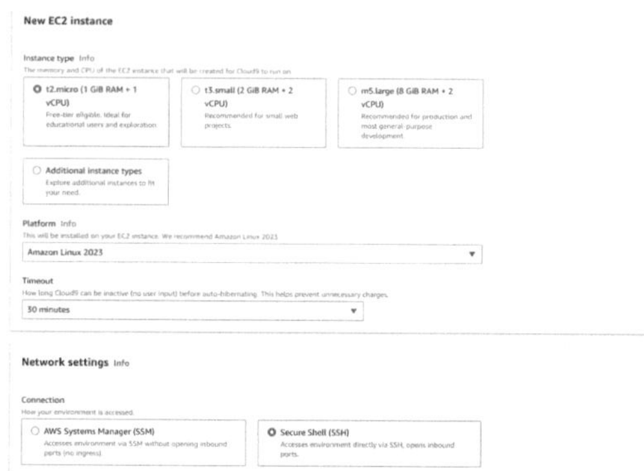


**FIGURE 3.3**
Cloud9 Creation Settings

There are over 200 services on AWS. Therefore, the *search bar* on top of the console is the quickest way to find a specific AWS service. We can type *Cloud9* in the search bar and open the Cloud9 service page from the result list. On the Cloud9 page, we need to create an *environment* on Cloud9. We give the environment a name and description on the environment creation page. We can also choose the *EC2 instance* type and define the *Network settings*, as shown in Figure 3.3. Powerful instances can execute codes faster but cost more credits. In addition, we can only select the available instance types and permitted network settings listed in the lab Readme file. In this section, we can choose the cheapest *t2.micro* instance type and the *Secure Shell* in the network settings.

Once the environment is ready, we can open the Cloud9 editor, as shown in Figure 3.4. On the left side, we can see the *folders* that contain different files in the environment, a *Git icon* to use Git Commands, and available *AWS resources*. The middle section shows a welcome page where we edit and execute Python codes. Beneath the welcome page, we can access the Linux instance terminal that hosts this environment. We can install Python libraries or execute Git commands in the terminal. On the right side, we can change the environment settings, chat with other developers, and debug our codes.

**FIGURE 3.4**
Interface of Cloud9 IDE

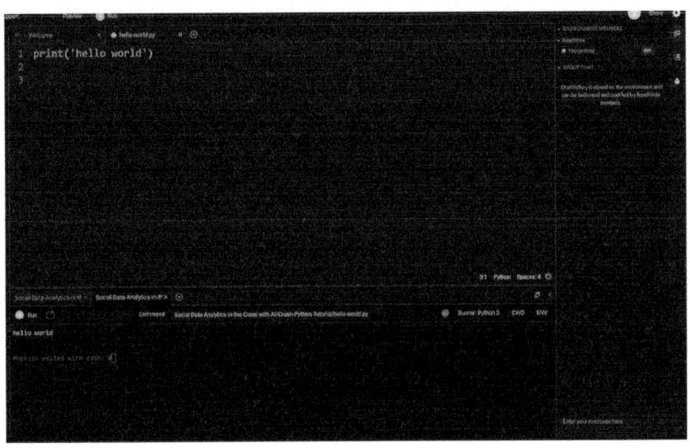


**FIGURE 3.5**
Output of the First Python Code

We are now ready to connect the Cloud9 editor to our GitHub repository. Click the *Git icon* on the left side of the Cloud9 editor, select *Clone Repository*, and provide the repository URL. For example, the URL of this textbook is https://github.com/xbwei/Social-Data-Analytics-in-the-Cloud-with-AI .git. Hit the *Enter* key, and the GitHub repository, including the license and README files, will be downloaded to the Cloud9 folder.

Right-click the repository folder and add another folder. We can name it *Crush-Python-Tutorial* and use this folder to store all the Python codes of this section. Using a dash line or underscore to separate words in folder or file names is also recommended. We can add a new file and name it *hello-world .py* within the folder. Double-click the hello-world.py file, and type our first Python code in the opened window:

```
print('hello world')
```

Click the *Run* button to execute the Python code, and you can see the output of 'hello world' being printed out, as shown in Figure 3.5.

Now we have a new Python file created in our Cloud9 Editor. We will synchronize our Cloud9 folders with the GitHub repository by uploading the newly created Python file to GitHub. We must first authorize Cloud9 to upload files to GitHub. Therefore, we must generate a *token,* like a password, on GitHub. We can then pass the token to Cloud9 so that Cloud9 can be authorized to communicate with the repository. We can open the GitHub token page[14] and generate a new token. The best practice for using a token is to rotate the token regularly, i.e., the token will expire in a few days or weeks, and a new token needs to be generated. However, considering we are academic users, we choose *no expiration* to simplify our work. For the token's scope, we should grant the least privilege, i.e., the exact permissions that are strictly required. In our case, the *repo* scope is sufficient. Figure 3.6 shows the token configuration where we provided a meaningful note, selected no expiration, and chose repo as the only scope. Once the token is created, we must copy it immediately and save it safely. For security purposes, the token will be hidden after a few minutes.

Back in Cloud9, we need to type the following git commands in the instance terminal:

```
git config --global credential.helper store.
git config --global user.email "you@example.com"
git config --global user.name "Your Name"
```

The first command tells Cloud9 to store tokens in a plain text file on the Cloud9 instance. Please note: Storing credentials without encryption is not recommended in production. AWS Identity and Access Management (IAM)[15] is the best solution for storing tokens on Cloud9. However, AWS IAM is not supported in the AWS Academy Learner Lab, so we can only store the token

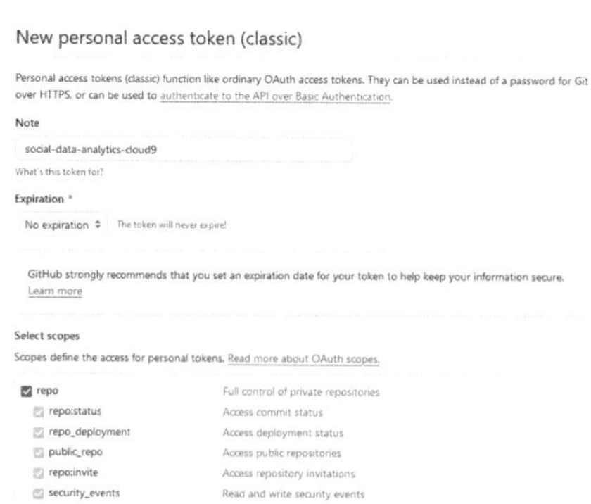


**FIGURE 3.6**
Generate GitHub Token

as plain text. The following two commands store the email and user name of the person who will execute Git commands, and you need to replace them with your GitHub account email and your name. This information will be recorded on the GitHub version history when we upload codes.

Next, we click the Git icon and see the new Python file is tracked as a change. As shown in Figure 3.7, we can click the + *icon* to *stage* all changes and leave a *comment* to explain the change, such as *our first python code.* We can then click the menu icon to *commit* the change. Finally, click *push* in the Git menu to complete the synchronization. In the following window, provide the GitHub user name and the created token (not the GitHub password). Our GitHub username and token will be saved on the Cloud9 instance, so we don't need to type the use name and token in future git commands.

Now, our *hello-world.py* has been uploaded to the GitHub repository. We can end the Learner Lab to save our AWS credits. The Cloud9 instance will be stopped automatically after 30 minutes of inactivity. We can also view the codes on the GitHub repository anytime, as it is a free service.

## Data Types and Containers

If you logged out of AWS Academy and ended the Learner Lab after the previous section, you can log back to AWS Academy, restart the Learner Lab, and open the Cloud9 editor. The Python codes and GitHub credentials are preserved. Next, we can create a new file under the *Crush-Python-Tutorial* folder and name it *data-types-and-containers.py.*

Like many other programming languages, we can write *comments* in Python to explain or describe the purposes of our codes. Complex Python codes are always hard to understand without comments, even for the authors of those codes. Therefore, providing clear and concise comments is a best practice in programming. In addition, the Python interpreter doesn't execute comments, so we can put a piece of code into comments when we debug our codes. We can use a pair of three quotation marks, either single or double

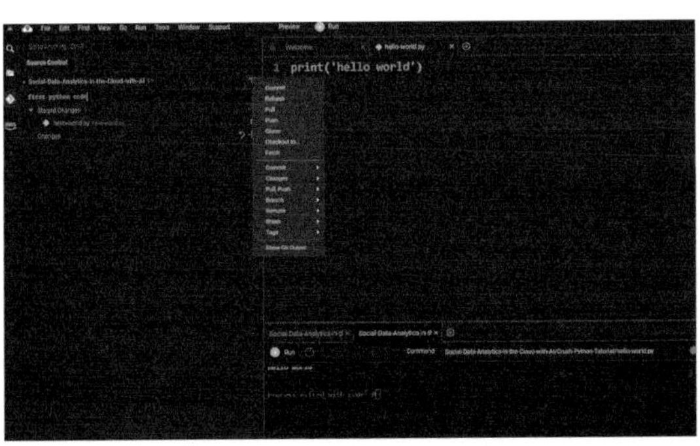


**FIGURE 3.7**
Commit Git Changes

quotations, to define a *regional comment* or a hashtag to define *a single-line comment*. For example, we can type the following codes in the *data-types-and-containers.py* file:

```
"""
data types and containers
"""
#Introduce the basic data types and containers in
Python
```

In the above example, we defined a regional comment to serve as the title of this Python file and used a single-line comment to describe the purpose of this file. The Python interpreter ignores all the comments, so no output will be printed when executing the codes.

There are three basic data types in Python: *string; integer;* floating-point numbers, or *float* for short.

- string: a series of characters, and is declared by using a pair of quotation marks, either single quotations or double quotations
- integer: whole numbers declared without quotation marks
- floating-point numbers: numbers with a decimal representing a certain level of precision, declared without quotation marks

We can use the *type* function to check data types in Python:

```
# basic data types

print(type("123"))
print(type(123))
print(type(123.))
```

The output shows that the first code defines a string because the value is declared in a pair of quotation marks. The second code defines an integer, and the third creates a floating-point number. We can tell the difference between an integer and a float by whether a decimal is presented in the number.

```
<class 'str'>
<class 'int'>
<class 'float'>
```

Each data type has specific *methods*, namely, actions that can be performed on a piece of data. We use dots and parentheses to tell Python to make the method act on the data and provide more information within the parentheses. For example, below are some standard string methods and manipulations:

```
# common string methods

print("Hello, World".upper()) #change to upper cases
print("Hello, World".lower()) #change to lower cases

print(" Hello, World ") #a string with spaces on both
sides
print(" Hello, World ".strip()) #strip spaces on both
sides

print("Hello, World".split()) #split the string by space
print("Hello, World".split(',')) #split the string based
on ,

print("Hello, " + "World") #concatenate two strings

print(len("Hellow, World")) #count the number of
characters
print("Hellow, World"[0]) #print the first character
```

The first two lines use the *upper* and *lower* methods to change a string to upper and lower cases. The *strip* method can delete the spaces at the beginning or end of a string. The *split* method can separate a string into substrings by spaces. If a delimiter is provided, such as the comma in the above example, the method splits the string based on the delimiter. The *len* function counts the number of characters in a string, and a space is counted as a character. We can also access a single character in a string by specifying its position in brackets. Please note: in Python, the position of the first element in a sequence is always 0, not 1. Below are the outputs of the above codes.

```
HELLO, WORLD
hello, world
Hello, World
Hello, World
['Hello,', 'World']
['Hello', ' World']
Hello, World
13
H
```

In Python, we can calculate summation, difference, produce, power, and division for numbers. Below are some examples:

```
# number calculation

print(2+3) #sum
print(2-3) #difference
print(2*3) #product
print(2**3) #power
print(2/3) #division
```

The output of those calculations is below. Users can check the Python document for advanced mathematical functions.[16]

```
5
-1
6
8
0.6666666666666666
```

In Python, each value is an *object*, and we can use *variables* to label or distinguish different objects. We can also change the value of a variable referred to at any time, and Python always keeps track of the current value.

```
# define variables

my_name = "Andrew"
print(my_name)

my_name = "Andy"
print(my_name)
```

In the above example, we defined a variable called *my_name*, which refers to the string *Andrew*. As shown below, when we print my_name, the output is *Andrew*. If we assign the variable to *Andy*, the value *Andy* is printed out.

```
Andrew
Andy
```

In Python, we can use formatted string literal, or *f-string*, to pass variable values or expressions into a string. We can start a string prefixed with *f* and enclose the variable or expression in curly braces, like the example below:

```
# f string
print(f"my name is {my_name}") #pass a variable into
string
print(f"{2*3}") #pass an expression into string
```

The output of the above code is:

```
my name is Andy
6
```

When we define a variable name, we want it to be meaningful and easy to understand. As a conversion, Python programmers use *underscores* to separate words in variable names for easy reading. A variable name can start with strings or underscores and contains numbers, strings, and underscores. However, you cannot use the reversed Python Keywords[17] as variable names.

Four types of data containers can store multiple values in Python: *list, tuple, set,* and *dictionary*. A list is a sequence of values in a particular *order*. It can be declared with square brackets or the *list function*. A list can store strings, numbers, variables, or other data containers and is *mutable*, i.e., we can reassign the items in a list. Like strings, we can access items in a list by calling the *index*, namely the position of the item. We can also make a list slice by specifying the index of the first and last items. Please note: In Python, the index starts at 0 from the left and -1 from the right. If we want to return the items with the index between n and m, we should define the slice starting at *n* and ending at *m+1*.

```
# define a list

my_list =[1,1,3,4] #define a list
print(my_list)
my_nested_list = [1,1,3,4,[1,2]] #define a nested list
print(my_nested_list)
print(my_list[1:3]) #print the 2nd and 3rd item
my_list[0]=0 #change the value of the first item
print(my_list)
```

The above codes define a list and a nested list, and the index is used to access the second and third items and change the value of the first item. The output of those codes is below.

```
[1, 1, 3, 4]
[1, 1, 3, 4, [1, 2]]
[1, 3]
[0, 1, 3, 4]
```

Once we define a list, we can *append* new items, *remove* unwanted items, *sort* them based on values, and *reverse* the order. The addition operator can combine multiple lists. We can count the number of items in a list with the len function or check whether an item exists with the *in* operator. Below are some examples of standard list methods.

```
#list methods

my_list.append(7) #add an item to a list
print(my_list)

my_list.remove(3) #remove an item from a list
print(my_list)

my_list.sort() #sort the items in a list
print(my_list)

my_list.reverse() #reverse the list order
print(my_list)

print(my_list+my_nested_list) #combine two lists
print(len(my_list)) #count the number of items in a
list
print(7 in my_list) #check if an item in a list
```

The output of the above codes looks like this:

```
[0, 1, 3, 4, 7]
[0, 1, 4, 7]
[0, 1, 4, 7]
[7, 4, 1, 0]
```

```
[7, 4, 1, 0, 1, 1, 3, 4, [1, 2]]
4
True
```

A set stores *unordered* items and is created with curly brackets or the *set function*. The items in a set must be *unique*, and all duplicated items will be removed. A tuple can also store multiple items and is declared with a *comma* or the *tuple function*. As a convention, programmers enclose a tuple in parentheses. However, a single value in parentheses without a comma is not a tuple. In addition, tuples are *immutable*; we can't reassign values to a tuple once defined.

```
# set and tuple

my_set = {1,1,3,4} #define a set
print(my_set) #duplicated items are removed
my_tuple = (1,) #define a tuple
print(type(my_tuple)) #this is a tuple
not_a_tuple = (1) #this is not a tuple
print(type(not_a_tuple))
# my_tuple[0] = 2 #this will create an error
```

In the above example, we defined a set with duplicated values; only distinct values are stored and printed out. We also defined a tuple in parentheses with a comma; the printed data type is a tuple. If we omit the comma, the printed data type is an integer. Finally, if we try to reassign a value in a tuple, we will hit an error. Below are the outputs:

```
{1, 3, 4}
<class 'tuple'>
<class 'int'>
```

Python dictionaries store *key-value pairs* and can be created with curly brackets or the *dict function*. Dictionary keys can be strings or numbers; dictionary values can be strings, numbers, variables, or other containers. To access a value in a dictionary, we can call the corresponding keys in a square bracket.

```
# define a dictionary
my_tweet = {
 "tweet_id":1138,
 "coordinates": (-75, 40),
 "visited_countries": ["GR", "CH", "MY"]
}
```

In the above example, *my_tweet* is a dictionary, the value of the *coordinates* key is a tuple, and the value of the *visited_countries* key is a list. In this case, we can access the nested containers by calling the dictionary keys and applying the associated methods to those values.

```
print(my_tweet['visited_countries']) #this equals to
the list of ["GR", "CH", "MY"]

print(len(my_tweet['visited_countries'])) #use len()
function to count the number in the list

print("US" in (my_tweet['visited_countries'])) #use
in operator to check whether 'US' in the list
```

For example, when we call the *visited_countries* key in the my_tweet dictionary, we get a list. We can then use the *len* function to count the number of items in that list or the *in* operator to check whether a specific value is in the list. The output is below.

```
['GR', 'CH', 'MY']
3
False
```

We can also print all the keys or values in a dictionary with the *keys* and *values* methods and add new key-value pairs as follows:

```
# dictionary methods

print(my_tweet.keys()) #print all the keys in a
dictionary
print(my_tweet.values()) #print all the values in a
dictionary
```

```
my_tweet['text']='hello world' #add a new key-value
pair
print(my_tweet)
```

The output will be like this:

```
dict_keys(['tweet_id', 'coordinates',
'visited_countries'])
dict_values([1138, (-75, 40), ['GR', 'CH', 'MY']])
{'tweet_id': 1138, 'coordinates': (-75, 40), 'visited_
countries': ['GR', 'CH', 'MY'], 'text': 'hello world'}
```

Table 3.1 summarizes those data containers in Python. Please refer to the recommended textbooks or the official document [18] for more about data types and containers.

Now, we have a new data-types-and-containers.py file. Please use the Git commands introduced in Figure 3.7 to upload the Python file to the GitHub repository. We no longer need to provide a username and token when uploading the codes because we have already stored them on the Cloud9 instance.

## Flow Control

As in previous sections, let's create a new file and name it *flow-control.py* under the *Crush-Python-Tutorial* folder.

**TABLE 3.1**

Summary of Python Data Containers

| Name | Purposes | Declaration |
|---|---|---|
| List | A collection of values in a particular order | Defined with a pair of square brackets or the list function |
| Tuple | A collection of immutable values | Defined with a comma or the tuple function, typically enclosed in a pair of parentheses |
| Set | An unordered collection of unique values | Defined with a pair of curly brackets or the set function |
| Dictionary | A collection of key-value pairs | Defined with a pair of curly brackets or the dict function |

In Python, *whitespaces* inside parentheses or brackets are ignored. We can use a *backslash* to start a new line when the code is too long. For example, the following codes are legit.

```
#whitescape
print(1
+ 22) #whitespaces in parentheses and brackets are
ignored
my_demo = 2+\
2 #use a backslash to continue a new line
print(my_demo)
```

As mentioned before, all values and data containers are objects in Python. Therefore, we can use the *id* function to check an object's identity. We can also use the *is* operator to compare the IDs of two objects and the *double equation* operator to compare their values.

```
# object id vs. value
list_a = [1,2,3]
list_b = [1,2,3]
print(id(list_a)) #get the id of a list variable
print(id(list_b))
print(list_a is list_b) #compare the ids of two
variables
print(list_a==list_b) #compare the values of two
variables
```

In the above example, we defined two variables that refer to the same list. By printing their IDs, we can see that two variables have different object IDs, even though they have the same values. The results of these codes are below:

```
140689067887872
140688840061632
False
True
```

In addition to the basic data types, there are two special data types in Python: *Boolean* and *none*. The Boolean data type has only two values, i.e., *true* and *false*. The Boolean data can be used with the *and*, *or*, and *not* operators. The none data type indicates a null value in Python and has only one value: none. In Python, empty data containers, empty strings, integer 0, float 0.0, and none are all considered *false*. Please test the following codes and see whether the results make sense.

```
# Boolean and None

print(False and True) #return False
print(False or True) #return True
print(not False) #return True

print(not []) #empty list is False, therefore not []
return True
print(not None) #None is False, therefore not None
return True
```

Flow control defines the order of computations in Python. The basic flow control methods include *if-else* statements, *for* loops, the *range* function, *while* loops, and *continue/break/pass* statements.

The if-else statements are used for conditional executions. Python executes the codes beneath the *if* statement when the conditional test evaluates true. The *else* statement is optional and allows you to define actions when the conditional test fails. Please note: indentations in Python tell the interpreter which lines are inside a flow control statement. Lines without indentations are always executed. The following is an example of using an if-else statement:

```
# if-else statement

if 2>1: # the condition is true
    print('2 is greater than 1') #this code will be
    executed

else:
    print('2 is less than 1') #this line will be
    ignored
print('not in the flow control') #this line will
always be printed as it is not inside an if-else
statement
```

The results of the above code are:

```
2 is greater than 1
not in the flow control
```

The *for* loop iterators over items of a sequence, such as data containers or sting values, in the order that they appear in the sequence. Below is a simple example of using a for loop to iterate items in a list. Please note: indentations are also added before the *print* function to indicate that the function is inside the loop.

```
for number in [1,2,3]:
    print(number)
```

The output of the above code is:

```
1
2
3
```

The for loop can be used with the *range* function, which generates arithmetic progressions. The range function has three arguments: the *start* value, the *end* value, and the *step*. It generates a progression from the start value to the *end value minus one*, with an increment of the step. The default start value is 0, and the default step is 1.

```
# for loop and range function

for number in range(1,7,2): #it will start from 1, end
at 6 (7-1) with an increment of 2
    print(number) #it will print 1, 3 (1+2), and 5 (1+2+2)
```

The above example will print numbers starting from 1 with an increment of 2. However, the number 7 will not be printed, as the maximal value of the range function is 6, i.e., 7-1. Here is the output:

```
1
3
5
```

The while loop can repeat executions as long as a condition is *true*. The condition should eventually be changed to *false* to avoid a dead loop. Below is a simple example of a while loop. Before you run the codes, try to figure out the output yourself.

```
# while loop

number = 1 #define the initial value to be 1

while number <=5: # condition is the number is less or
equal to 5

    number = number + 1 # change the number value to
    avoid a dead loop

    print(number)
```

Here is the output:

```
2
3
4
5
6
```

The *break* statement can break out of *the smallest* enclosing *for* or *while* loop; the *continue* statement *skips the current interaction* and continues with the next iteration; and the *pass* statement is used when a statement is syntactically required, but *no action* is needed. The following example uses the *while* loop to print numbers from 2 to 6. Try switching the *continue* to the *pass* or *break* statement to see how the results change. Please note: You may create a dead loop.

```
# while loop and break/continue/pass statement
number = 1
while number <=5:
    number = number + 1
    if number ==3:
        continue # try to use pass or break to see what
        will happen
    print(number)
```

Flow controls make programming powerful and are used in every piece of coding. Please check the recommended textbooks or the official Python document[19] for more about flow control tools. Please also remember to upload the *flow-control.py* file to your GitHub repository.

## Program Design

Let's create a *program-design.py* file under the *Crush-Python-Tutorial* folder. You can upload this file to GitHub after completing this section.

As programmers, we consistently generate errors, either intentionally or involuntarily. In Python, errors detected during executions are called *exceptions*, and we can use the *try* and *except* clause to handle those exceptions. We put our "normal" codes in the try clause and the solutions to handle exceptions in the except clause. If no exceptions occur, the except clause will be skipped. Otherwise, the solutions in the except clause will be invoked.

```
# try-except clause

for number in range(5):
    try:
            print(1/(3-number))
    except:
            print('error')
            pass
```

In the above example, we use the *range* function to generate numbers from 0 to 4 and then bring each number into the division calculation. An error will be created when the number equals 3, as the division calculation would be 1 divided by 0. When that error is detected in the *try* clause, Python will execute the codes in the *except* clause, namely print the 'error' string and continue to the next iteration. The remaining iterations will continue until the end or reach another exception. Below is the output of the above codes:

```
0.3333333333333333
0.5
1.0
error
-1.0
```

We can also define multiple except clauses following a single try clause to handle various errors. For example, the error in the above example is a

*ZeroDivisionError,* and we can specify the error type in the except clause to provide different solutions. For other types of error handling, please check the official Python document.[20]

Most editors have a *debugging* tool to help programmers understand complex codes or track errors. In Cloud9, we can right-click the line number of a code and choose *Add Breakpoint.* A red dot will appear, and the interpreter will stop at those breakpoints when the codes are executed. We can also check the current values of all variables in the debugging window to understand better how the codes work. As shown in Figure 3.8, a breakpoint is added in *line 9.* When debugging mode is enabled, the interpreter stops at line 9, and the current value of the *number* variable is reported on the right side. We can execute the code step by step to closely monitor the changes in the *number* variable and quickly identify why an error occurs.

As our codes become complicated, we may see scenarios where we use similar codes multiple times. In that case, we can pack those codes in a *function* and reuse that function instead of repeating the same codes. Using functions makes our codes easier to write, read, and test. The following example defines a function to calculate the absolute values of a given input:

```
# define a function

def calcuate _ abs(input _ value):
    if type(input _ value) is str:
        return('wrong data type')
    elif input _ value>=0:
        return(input _ value)
    else:
        return(-input _ value)
```

In this example, we name the function *calculate_abs* and provide one argument, namely *input_value.* Within the function, we first test the data type of an input value and return the "wrong data type" if the value is a string. A

**FIGURE 3.8**
Debugging in Cloud9

function stops when a *return* clause is executed. If the input data type is not string, the function will return the absolute value of the input data. To use that function, we need to call the function name and provide the required arguments:

```
print(calcuate _ abs("a"))
print(calcuate _ abs(9))
```

The output of those lines is:

```
wrong data type
9
```

Python is an *object-oriented* programming language, meaning that we can define customized objects to organize functions or data. To use objects in Python, we first need to create *classes*, i.e., templates of objects. We can then create an object from the defined class called an *instance*.

```
# class and object
class car: # define a class
    maker = '' #define an attribute of the class
    def report_maker(self): #define a method, i.e., a
    function, of the class
        return(self.maker) #this method will return
        the maker attribute
my_car = car() #create an instance
my_car.maker = 'Tesla' #define the maker to be Tesla

print(my_car.report_maker()) #call the report_maker
method
```

The above codes are a simple example of defining a *car* class. The *car* class has a *maker* attribute and a *report_maker* method. The maker attribute can store a car object's information, and the report_maker method can print that information. A *my_car* object is created based on the *car* class, and the *maker* attribute is defined as "Tesla." When *my_car* object calls the *report_maker* method, "Tesla" is printed.

There is a rich collection of Python libraries in data analytics. To use those libraries, we first need to use the *pip* tool to install them in the terminal.

Please check the official Python document for more about installing Python resources.[21] We can then *import* those libraries into our codes. For example, we can pip install the NumPy[22] library in the terminal:

```
pip install numpy
```

We can then import the numpy library to our codes and use the *help* function to view the primary usage of the imported library.

```
# import a library

import numpy #you need to pip install the numpy in the
terminal first
help(numpy) #this line can be very slow
```

Bonus: Try to import *this* library and see what will happen!

```
# the Zen of Python

import this #see what will happen
```

It is uncommon to write classes in data analytics, and we can leverage AI to write complicated functions for us. Therefore, this section introduces the basics of designing classes and functions. Please read the two recommended textbooks for more about program design (Downey 2015; Matthes 2019).

## Python Programming with AI

GitHub Copilot[23] and Amazon Q Developer[24] are two AI-powered assistants that help write code in Python and other popular programming languages. Both tools can generate code based on users' comments and provide recommendations on existing codes. This section demonstrates Amazon Q Developer, which is available in the AWS Learner Lab. As we write this textbook, Amazon Q Developer is still called *CodeWhisperer,* the old name of Amazon Q Developer in Cloud9.

Let's create a new file named programming-ai.py under the Crush-Python-Tutorial folder. To enable CodeWhisperer in Cloud9, we open the *AWS Resource* on the left side of Cloud9 and *Resume the Auto-Suggestions* under CodeWhisperer, as shown in Figure 3.9.

Once the *Auto-Suggestions* are enabled, we can ask AI to write Python codes by providing instructions in comments. For example, we can write the following comment in natural language:

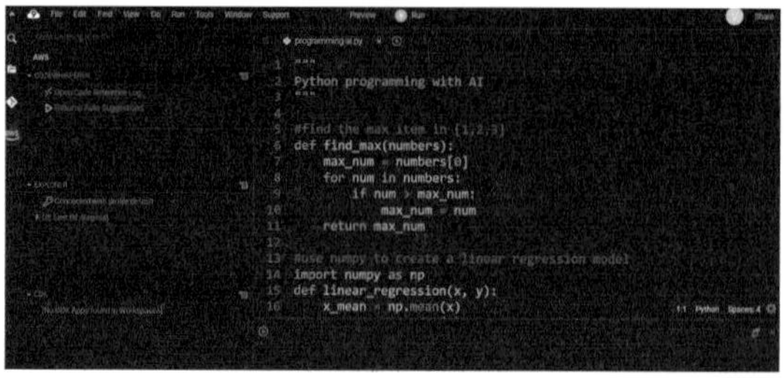

**FIGURE 3.9**
Enable CodeWhisperer in Cloud9

```
#find the max item in [1,2,3]
```

We then press the *Enter* key, and the following codes are automatically generated:

```
def find_max(numbers):
    max_num = numbers[0]
    for num in numbers:
        if num > max_num:
            max_num = num
    return max_num
```

CodeWhisperer creates a function named *find_max* and uses a for loop to iterate all the items in a list. During the iteration, each item is compared against a variable named *max_num*. The initial value of the max_num is the first item in the list. If a bigger number is found during the iteration, the max_num will refer to the bigger number. After the iteration, the max_num refers to the maximal item in that list. This function works for any list of number values.

If we keep pressing the enter key, CodeWhisperer can generate similar but useless codes, such as:

```
#find the min item in [1,2,3]
def find_min(numbers):
    min_num = numbers[0]
    for num in numbers:
        if num < min_num:
            min_num = num
        return min_num
```

In the above example, CodeWhisperer generated a similar comment to what we initially typed and completed the corresponding codes automatically.

We can ask CodeWhisperer to write more complicated codes. For example, in the previous section, we installed *numpy*, a robust scientific computing library. But we know nothing about numpy usage. We can ask CodeWhisperer "what numpy can do" in a comment, and CodeWhisperer can also answer that question in a comment. The human-typed question and the CodeWhisperer response are printed below:

```
#what numpy can do?
#numpy can do matrix multiplication, linear algebra,
and random number generation
```

If we keep pressing the Enter key, CodeWhisperer even provides two simple samples with comments:

```
#use numpy to multiply two matrices
import numpy as np
def matrix_multiply(matrix1, matrix2):
    return np.matmul(matrix1, matrix2)

#use numpy to find the mean of a list of numbers
def mean(numbers):
    return sum(numbers) / len(numbers)
```

In the above example, CodeWhisperer first provided a comment to explain the purpose of the sample code and then imported the numpy library as *np*, a standard convention in data analytics. It then defined a function to implement the task described in the first comment, i.e., multiplying two matrices. CodeWhisperer then provided a second example with a comment and sample code to find the mean of a given list. However, the *import* line is omitted in the second code since it is unnecessary.

We can try to ask complicated questions, such as "create a linear regression model," like the following:

```
#use numpy to create a linear regression model
```

CodeWhisperer generated the following code in a second. Please feel free to test the following function with two lists of numbers and see whether the result is correct.

```
import numpy as np
def linear_regression(x, y):
    x_mean = np.mean(x)
    y_mean = np.mean(y)
    num = 0
    den = 0
    for i in range(len(x)):
        num += (x[i] - x_mean) * (y[i] - y_mean)
```

```
            den += (x[i] - x_mean) ** 2
    m = num / den
    c = y_mean - m * x_mean
    return m, c
```

We can also ask CodeWhisperer to process datasets. For example, we can try the comment "upload an image to s3" and get the following responses:

```
#upload an image to s3
import boto3

s3 = boto3.client('s3')

def upload_image(image_path, bucket_name,
object_name=None):
    if object_name is None:
          object_name = image_path.split('/')[-1]
    s3.upload_file(image_path, bucket_name,
    object_name)
    return True
```

S3 is Amazon's cloud object storage. The above codes imported the *boto3* library, AWS's Python software development kit (SDK), and created a function to upload a local image to an S3 bucket.

Here is another example of asking CodeWhisperer to clean and analyze Twitter data with the popular OpenAI model:

```
#use openai to identify the sentiment of tweet data,
remove the urls and special characters in the tweets

import openai
import re

def clean_text(text):
    text = re.sub(r'http\S+', '', text)
    text = re.sub(r'@\S+', '', text)
    text = re.sub(r'#\S+', '', text)
    text = re.sub(r'[^\w\s]', '', text)
    return text
def get_sentiment(text):
    text = clean_text(text)
    response = openai.Completion.create(
          engine="text-davinci-002",
          prompt="Identify the sentiment of the
          following text: " + text,
          max_tokens=1024,
```

```
            n=1,
            stop=None,
            temperature=0.5,
        )
        return response.choices[0].text.strip()
```

In the above example, CodeWhisperer understands the syntax and usage of the OpenAI library, selects the appropriate model engine and token size, and completes the prompt in the function.

With those AI-powered assistants, we don't need to remember details of Python syntax and the usage of various libraries. However, a basic understanding of Python programming is essential, and we should be able to verify AI-generated codes as they may contain errors.

## Data Analytics with Python

Data analytics involves four components: data *collection, storage, process and query,* and *visualization.* Figure 3.10 shows how those components are connected. To start, analysts need data to investigate. The data may already be available in analysts' institutions, such as private data, or it may need to be purchased or collected from other resources, like commercial or open-source data. The collected data should be stored in a dedicated storage space to support quick analysis. Chapter 4 describes various data storage solutions and introduces how to collect data online.

The collected data then has to be processed. For example, the data may contain errors or outliers that must be removed. Analysts may also need to

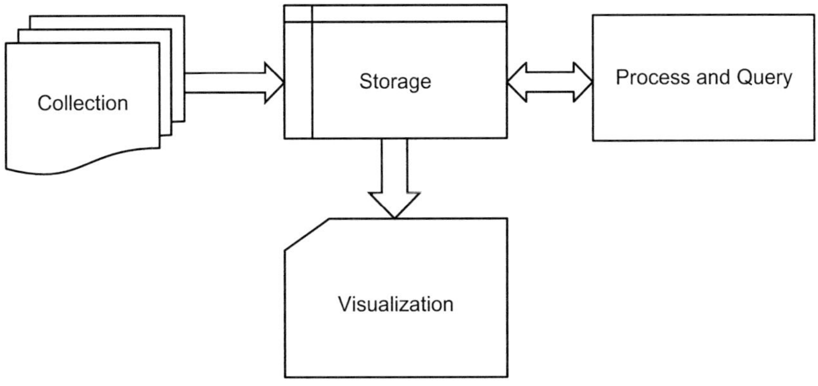


**FIGURE 3.10**
Data Analytics Components

aggregate and combine data from multiple resources, change the data formats, or transform the data into specific shapes for further analysis. Analysts can then query the data by applying statistical models or business logic to answer particular questions. Chapter 5 introduces various query languages and techniques to process and query data.

Data visualization is also essential in data analysis, as some patterns or insights can only be identified through visualizations. Chapter 6 presents the basic principles and best practices in data visualization and demonstrates how to create charts and dashboards with various tools.

A rich collection of Python packages is used in data analytics and data science. For example, *pandas*[25] provides an efficient *DataFrame* for data manipulation, including loading and exporting data, joining, cleaning, and reshaping data. *NumPy*[26] is fast in scientific computing, including mathematical and logical operations, shape manipulation, linear algebra, and statistical operations. *Matplotlib*[27] is frequently used in data visualization, and the output can be embedded in Jupyter Notebook. In addition, Scikit-learn[28] provides simple and efficient tools for "shallow" machine learning, while TensorFlow[29] and PyTorch[30] are suitable for deep-learning projects.

This section introduces basic data analytics with Python, including setting up JupyterLab, manipulating data with pandas, and simple data visualizations with Matplotlib.

## Set up JupyterLab

JupyterLab,[31] the next generation of Jupyter Notebook, is an open-source web application that can create and share documents with live codes, visualizations, and narrative text. The narrative text follows Markdown[32] syntax, providing formatted text with colors, styles, sizes, and hyperlinks. We can install a JupyterLab application locally or with a cloud service like Amazon SageMaker,[33] GitHub Codespaces,[34] and Google Colab.[35] This section teaches how to set up a JupyterLab on SageMaker and connect the notebook instance with a GitHub repository.

If you still keep the GitHub token created in the previous section, please use that token. Otherwise, you can create a new token on GitHub.[36] Next, we start the AWS Learner Lab and search the *SageMaker* service on the AWS Console. On the SageMaker page, click *Notebooks* on the left side to add a *Git Repository* and a *Notebook instance*. First, we click the *Git Repositories* and *Add Repository*. On the next page, we choose *GitHub* as the repository type and provide a name and the URL of our GitHub repository created in the previous section. We must also create a new secret to store the GitHub username and token. Figure 3.11 shows the settings for adding a GitHub repository on SageMaker.

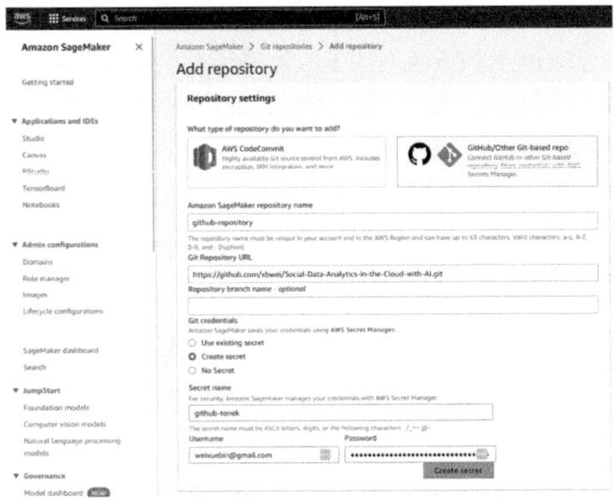


**FIGURE 3.11**
Add a GitHub Repository on SageMaker

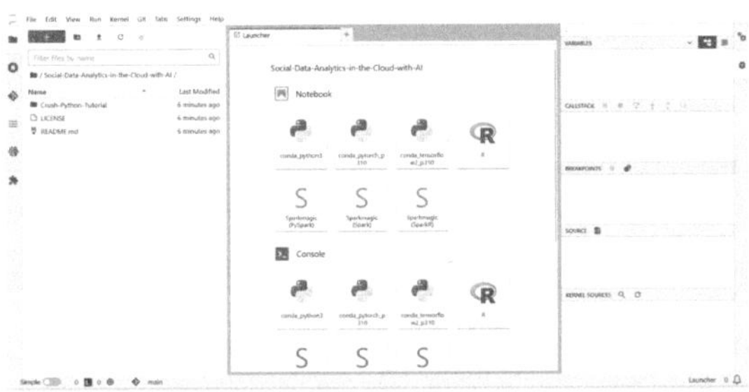


**FIGURE 3.12**
JupyterLab Interface

Next, we open the Notebooks page and *Create a notebook instance*. We can give the instance a name and accept the default settings, except to choose the created GitHub repository in the *Git repositories* option.

The Notebook instance may take a few minutes to be in service. We can then *Open JupyterLab*, where the GitHub repository files are already cloned to the Notebook instance. Figure 3.12 shows the interface of JupyterLab. Like Cloud9, JupyterLab has three sections: the left section displays the files on the instance. We can manage folders and files, run the Git commands, view sample notebooks provided by SageMaker, and install extensions on JupyterLab. We can edit a notebook or Python code in the middle section and debug our codes in the right section.

We can right-click the folder and create a notebook. Name the notebook *COVID_19_Data_Analysis.ipynb* and select the *conda_python3* kernel. Notebooks organize codes by *cells*, and we can write a piece of Python or Markdown code in each cell and execute those cells separately. By clicking the inside of a cell, we are in the *Edit* mode and can edit the content inside a cell. We can switch to editing Python *Code* or *Markdown* from the dropdown list at the top of the notebook. After editing, we can click any place outside a cell to enter the *Command* mode. In the command mode, we can select a cell, click the *Run* button to execute it, or drag and drop cells to change their orders. Please refer to the official JupyterLab document[37] for information on navigating notebooks.

The Markdown documents provide additional descriptions of the Python codes and can be written in formatted texts. Below are some essential Markdown elements:

- Headings: To indicate the level one heading, use a single hashtag before a line; two hashtags indicate level two, etc.
- Emphasis: Add one underscore before and after a word to add a bold effect and two underscores for an italic effect.
- Lists: Add numbers followed by periods in front of each line to define numbered lists. Use dashes to create unnumbered lists.
- Links: Write the link text in brackets followed by parentheses that contain the link URL.

For example, we can write the following Markdown code in the first cell as the title of the notebook:

```
# A Simple Example of Analyzing COVID-19 Data
```

We can add the data source with its website URL in the second cell, such as:

```
## Data source
The data is downloaded from [European Centre for
Disease Prevention and Control](https://www.ecdc
.europa.eu/en/publications-data/download-todays-data
-geographic-distribution-covid-19-cases-worldwide)
```

To learn more about Markdown syntax, please check the Markdown guide.[38] Readers can also visit the textbook's GitHub repository to view the *COVID_19_Data_Analysis* notebook.[39]

## Manipulate Data

This section provides a simple example of using *pandas* to load and manage data on worldwide COVID-19 cases. You can download the COVID-19 data from the *European Centre for Disease Prevention and Control.*[40] The downloaded data can be uploaded to the JupyterLab instance or stored in a separate data

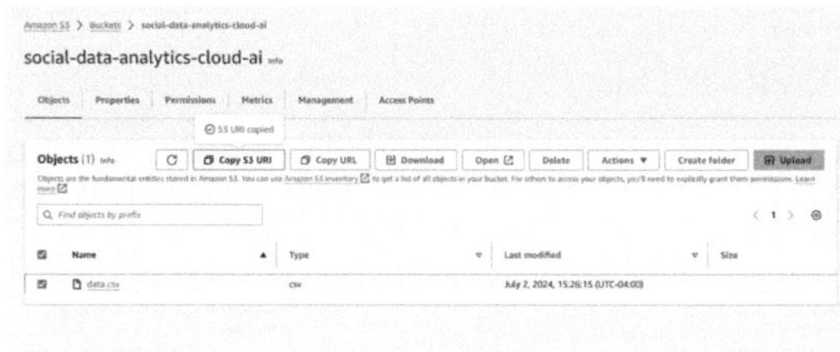


**FIGURE 3.13**
COVID-19 Data in an S3 Bucket

container. The best practice is to separate data computing from data storage for better efficiency. Therefore, we store the data on Amazon S3,[41] a cloud object storage service.

First, we need to create an Amazon S3 bucket, an object container that can store structured, semi-structured, and unstructured data such as images, videos, CSV files, etc. Search *S3* on the AWS console and open the Amazon S3 service page. We then *Create a new bucket* on the S3 service page, provide a name, and accept all the default settings. Next, we upload the COVID-19 data to the created bucket and *Copy the S3 URI* of the uploaded data, as shown in Figure 3.13. S3 URI stands for S3 unique resource identifier and is a convenient way for other AWS services, such as SageMaker, to access data on S3. By default, the data stored in an S3 bucket is private and can only be accessed by other services under the same AWS account. It is also possible to set the data public.

Pandas can load data from different types of files or resources, including comma-separated values (CSV) files, Excel files, HTML tables, or a query in a database.[42] It passes the loaded data into a *DataFrame* object, the most versatile and widely used data structure.[43] Because our data is a CSV file, we use the pandas' *read_csv* function to load it directly from S3.

```
import pandas as pd
df = pd.read_csv('s3://social-data-analytics-cloud-ai/
data.csv') # load the data from s3
df.head() #show the top 5 rows
```

The above codes load the COVID-19 data from the S3 bucket with the S3 URI and pass the data to a DataFrame object called *df*. We then use the *head* function of the DataFrame object to show the top five rows. The output is printed beneath the cell. The file contains the number of COVID-19 cases and deaths per day in each country. The country's population and cumulative case numbers are also included. Table 3.2 shows parts of the output. Please check the textbook's GitHub repository to view the whole table.[44]

**TABLE 3.2**

A Sample of the COVID-19 Data

| | dateRep | day | month | year | cases | deaths |
|---|---|---|---|---|---|---|
| **0** | 14/12/2020 | 14 | 12 | 2020 | 746 | 6 |
| **1** | 13/12/2020 | 13 | 12 | 2020 | 298 | 9 |
| **2** | 12/12/2020 | 12 | 12 | 2020 | 113 | 11 |
| **3** | 11/12/2020 | 11 | 12 | 2020 | 63 | 10 |
| **4** | 10/12/2020 | 10 | 12 | 2020 | 202 | 16 |

DataFrame is a two-dimensional data structure, and we can check the number of rows and columns by using the *shape* attribute:

```
df.shape #show the number of rows and columns
```

The output indicates that there are 619,000 rows and 12 columns:

```
(61900, 12)
```

When accessing *columns* in DataFrame, we can treat it as a Python *dictionary*. For example, to select the values in a single column, we can call the column name in a bracket:

```
df['cases'] #select a column
```

The above code returns all the values in the death column, as printed below:

```
0           746
1           298
2           113
3            63
4           202
            ...
61895            0
61896            0
61897            0
61898            1
61899            1
Name: cases, Length: 61900, dtype: int64
```

We can sort the values of a single column using the *sort_values* function. The following code sorts the entire table based on the number of *deaths* in ascending order. Please check the textbook's GitHub repository to view the result.[45]

```
df.sort_values(by='deaths') #sort data by a column
```

To access *rows* in a DataFrame, we can treat it as a Python *list* and slice the DataFrame with an *index*. For example, the following code returns the top 5 rows in the data, and the output is similar to the *head* function.

```
df[0:5] #slice by rows
```

We can also filter rows by values. For example, we can report all the rows where the number of deaths exceeds five. The following code returns 13,970 rows where the number of deaths exceeds five. Please check the pandas' document for more information about indexing and filtering data.[46]

```
df[df['deaths']>5] # select rows by values
```

We can also use pandas to explore and clean data. For example, the *info* function lists the data types and the number of null values for each column:

```
df.info() #show the data types and number of null
values
```

In the output, the *dateRep* column is not in the *date* format but an *object* format:

```
RangeIndex: 61900 entries, 0 to 61899
Data columns (total 12 columns):
#  Column   Non-Null   Count      Dtype
---         ------     ---------  -----  -----
0  dateRep  61900      non-null   object
1  day      61900      non-null   int64
2  month    61900      non-null   int64
3  year     61900      non-null   int64
4  cases    61900      non-null   int64
5  deaths   61900      non-null   int64
dtypes: float64(2),    int64(5),  object(5)
```

We can thus convert the *dateRep* data type to date for accurate data analysis and visualization:

```
df['date'] = pd.to_datetime(df['dateRep'],
format='%d/%m/%Y') #change the data type to date
with specific format
```

We can rename the columns to short ones, like calling the *popData2019* column *pop* and the *Cumulative_number_for_14_days_of_COVID-19_cases_per_100000* column as *cum*.

```
df = df.rename(columns={"popData2019": "pop", "Cumu
lative_number_for_14_days_of_COVID-19_cases_per_10
0000": "cum","countriesAndTerritories":"country" })
#rename columns
```

The *isnull* function can help identify the number of null values in each column. The following code calculates the number of null values in each column.

```
df.isnull().sum() #show the number of rows
containing null value
```

In the output, we can see the *pop* and *cum* columns have many null values:

```
dateRep 0
day     0
month   0
...
Cases   0
Deaths  0
Country 0
Pop     123
Cum     2879
Date    0
dtype: int64
```

We can then drop those null values by using the *dropna* function. The below code drops all the rows with null values.

```
df = df.dropna() #drop null values
```

The *describe* function reports a statistical summary of the data, including the mean, minimal, maximal, lower and upper percentiles, and standard deviation. Those statistics help quickly identify the patterns and potential errors in the data.

```
df.describe() #statistical summary of the data
```

**TABLE 3.3**

The Statistical Summary of the COVID-19 Data

| | cases | deaths | pop | cum | date |
|---|---|---|---|---|---|
| **count** | 58759.000000 | 58759.000000 | 5.875900e+04 | 58759.000000 | 58759 |
| **mean** | 1216.417179 | 27.441890 | 4.138481e+07 | 66.470760 | 2020-07-24 |
| **min** | -8261.000000 | -1918.000000 | 8.150000e+02 | -147.419587 | 2020-01-13 |
| **25%** | 0.000000 | 0.000000 | 1.293120e+06 | 0.762010 | 2020-05-17 |
| **50%** | 20.000000 | 0.000000 | 7.813207e+06 | 6.698860 | 2020-07-27 |
| **75%** | 312.000000 | 5.000000 | 2.860872e+07 | 52.618251 | 2020-10-05 |
| **max** | 234633.000000 | 4928.000000 | 1.433784e+09 | 1900.836210 | 2020-12-14 |
| **std** | 6952.740171 | 134.548086 | 1.540087e+08 | 162.657960 | NaN |

After reading the statistical summary in Table 3.3, we find negative values in the number of deaths and cases.

Those might be errors, so we can filter the rows that have negative deaths and cases:

```
df = df[(df['deaths']>=0) & (df['cases']>=0) &
(df['cum']>=0)] #filter data
```

We can use the *value_counts* function to count the number of unique values in a column, such as number of records in different countries:

```
df['country'].value_counts() #show the count of
unique values
```

The output shows that there are roughly 337 records for most countries, but some countries, like *Vanuatu*, have less than 50 records. We could remove countries with fewer records if necessary.

```
country
Japan                337
Iran                  337
Brazil               337
Russia               337
Switzerland          337
...
Comoros              214
Lesotho              201
Solomon_Islands      47
Marshall_Islands     34
Vanuatu              21
Name: count, Length: 211, dtype: int64
```

**TABLE 3.4**

Filtered Records by Using Index

| date | country | cases | deaths | pop | cum |
|---|---|---|---|---|---|
| **2020-04-05** | Afghanistan | 35 | 1 | 38041757.0 | 0.646658 |
| **2020-04-05** | Albania | 29 | 2 | 2862427.0 | 8.978395 |
| **2020-04-05** | Algeria | 80 | 25 | 43053054.0 | 2.582860 |
| **2020-04-05** | Andorra | 27 | 1 | 76177.0 | 496.212768 |
| **2020-04-05** | Angola | 2 | 0 | 31825299.0 | 0.025137 |
| **...** | ... | ... | ... | ... | ... |
| **2020-04-05** | Uzbekistan | 39 | 0 | 32981715.0 | 0.706452 |
| **2020-04-05** | Venezuela | 0 | 2 | 28515829.0 | 0.378737 |
| **2020-04-05** | Vietnam | 1 | 0 | 96462108.0 | 0.151355 |
| **2020-04-05** | Zambia | 0 | 0 | 17861034.0 | 0.207155 |
| **2020-04-05** | Zimbabwe | 0 | 0 | 14645473.0 | 0.047796 |

Not all columns store meaningful information. For example, we don't need columns such as *geoId* or *continentExp*, so we can keep the columns we are interested in:

```
df = df[['country','cases','deaths','pop','cum','d
ate']] # keep specific columns
```

We can set the DataFrame *index* to a specific column and filter values or merge tables based on that index. For example, we can set the *date* column as the index:

```
df.set_index('date',inplace=True) #define the index
```

We can then filter the rows by specific dates with the *loc* function:

```
df.loc['2020-04-05'] #filter the data based on the
index value
```

Table 3.4 reports the filtered records by using the index. The *date* column is now the first column as it is the index, and all the records from the day of *2020-04-05* have been returned.

Data in a DataFrame can also be exported to a file or stored in a database. The following code exports the cleaned data to a CSV file in the Notebook instance for reuse. We can also upload the data to the S3 bucket if needed.

```
df.to_csv('clean_data.csv') #store the cleaned data to
local computer
```

## Visualize Data

Pandas and Matplotlib are popular Python libraries for data analysis and visualization. When data is loaded in a pandas DataFrame, we can process and clean it, aggregate it depending on the questions we are interested in, and visualize the results with various charts. In addition, the *DataFrame.plot* is a convenient wrapper of Matplotlib, so we can use the plot function in DataFrame to create charts directly without explicitly importing the Matplotlib library.

To start, we can display the histogram of all columns:

```
df.hist(layout=(2,2), figsize=(10, 5), sharey=False,
sharex=False, bins=50) #use multiple charts to show
the distribution of all columns
```

Figure 3.14 shows that all columns have long right tails, indicating more outliers.

Next, we want to see the trend of cases over time. We first need to aggregate the results:

```
sum_cases_by_date = df.groupby('date')['cases'].
sum() #aggregate the total case number per day
```

The above example uses the *groupby* function to calculate the total number of COVID-19 cases per day and stores the results in a new DataFrame object. We can then visualize the result on a line chart:

```
sum_cases_by_date.plot() #use a line chart to show
the result
```

The result, as shown in Figure 3.15, indicates that fewer cases were reported in early 2020, and the number of cases increased dramatically and peaked at the end of 2020.

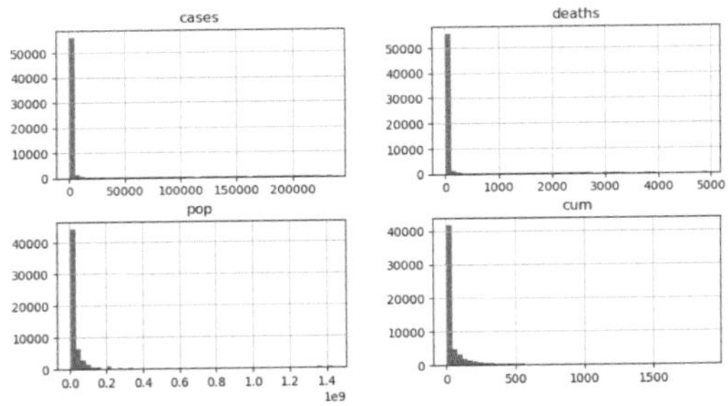


**FIGURE 3.14**
Histograms of COVID-19 Data

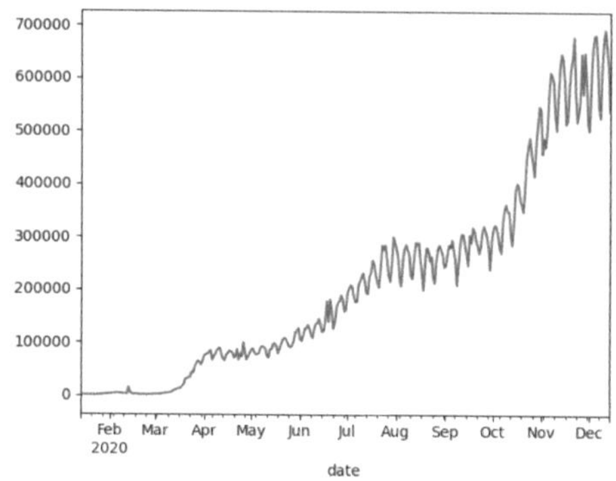


**FIGURE 3.15**
The Number of COVID-19 Cases per Day

We can also calculate the total deaths per country, identify the top 10 countries with the highest number of fatalities, and visualize the results with a bar chart:

```
sum_deaths_by_country = df.groupby('country')
['deaths'].sum() # calculate the total deaths per day
sum_deaths_by_country.nlargest(10).plot.bar() # use a
bar chart to show the results of the top 10 countries
```

The result is reported in Figure 3.16. It turns out that the United States has the highest number of deaths. So it might be interesting to explore the USA data only:

```
USA_data = df[df['country']=='United_States_of_Amer
ica'] #select the USA data
```

We can then visualize the number of deaths over time, where the colors represent the accumulated cases per 100,000 persons:

```
USA_data.reset_index().plot.scatter(x='date',y='de
aths',c='cum') #use scatter plot to show the result
```

The result is displayed in Figure 3.17. It shows that in May 2020, there were more deaths but fewer cases. Both numbers dropped in July and then peaked in December.

Please explore another country's data and remember to upload the notebook and data to GitHub. Once the notebook is uploaded, we can close JupyterLab, stop the Notebook instance, and end the Learner Lab to save credits.

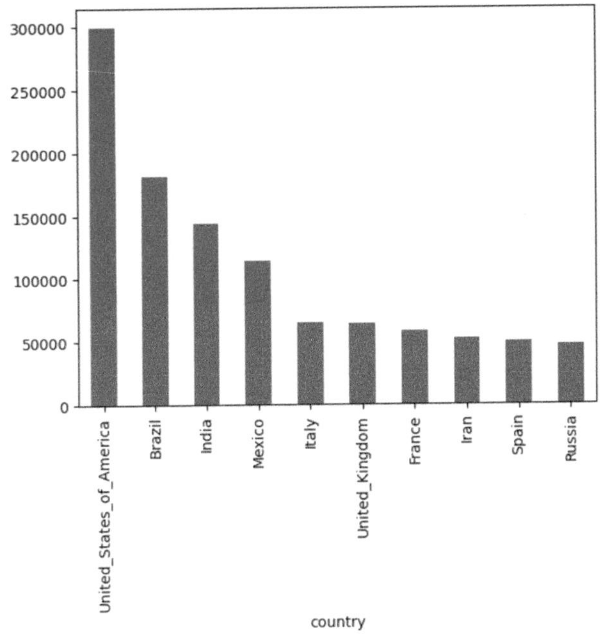


**FIGURE 3.16**
The Top 10 Countries with the Highest Covid-19 Deaths in 2020

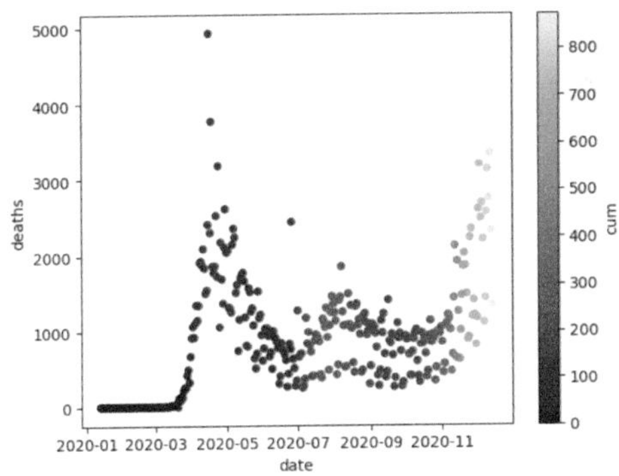


**FIGURE 3.17**
The Number of USA COVID-19 Deaths Over Time

This chapter offers a concise introduction to Python programming and illustrates how to analyze and visualize data using popular Python libraries. Readers should learn the fundamentals of Python syntax and structure, and then progress to practical applications involving data analysis and visualization tools. By the end of the chapter, readers should have a basic understanding of how to leverage these powerful libraries to manipulate datasets and create insightful visualizations. All the demo codes are available on the textbook's GitHub repository,[47] and additional learning resources are listed in the reference section. With AI assistants, developers and analysts no longer need to remember every detail of Python coding, allowing them to spend more time on application design or addressing research questions. These AI tools can provide code suggestions, debug errors, and offer real-time guidance, thereby streamlining the coding process. This enables professionals to focus on higher-level tasks and creative problem-solving, enhancing productivity and innovation in their projects. In Chapter 6, we will revisit data analytics in Python with the help of Generative AI.

## Notes

1. https://spectrum.ieee.org/the-top-programming-languages-2023
2. https://aws.amazon.com/pm/cloud9/
3. https://aws.amazon.com/q/developer/
4. https://jupyter.org/
5. https://www.python.org/
6. https://www.python.org/downloads/release/python-3124/
7. https://docs.python.org/3/library/idle.html
8. https://github.com/xbwei/Social-Data-Analytics-in-the-Cloud-with-AI
9. https://github.com/
10. https://github.com/new
11. https://opensource.org/license/mit
12. https://choosealicense.com/
13. https://www.awsacademy.com/vforcesite/LMS_Login
14. https://github.com/settings/tokens
15. https://aws.amazon.com/iam/
16. https://docs.python.org/3/library/math.html
17. https://docs.python.org/3/reference/lexical_analysis.html#keywords
18. https://docs.python.org/3/library/stdtypes.html
19. https://docs.python.org/3/tutorial/controlflow.html
20. https://docs.python.org/3/library/exceptions.html
21. https://packaging.python.org/en/latest/tutorials/installing-packages/
22. https://numpy.org/
23. https://github.com/features/copilot
24. https://aws.amazon.com/q/developer/
25. https://pandas.pydata.org/
26. https://numpy.org/
27. https://matplotlib.org/
28. https://scikit-learn.org/
29. https://www.tensorflow.org/
30. https://pytorch.org/
31. https://jupyter.org/
32. https://www.markdownguide.org/

33. https://aws.amazon.com/sagemaker/
34. https://github.com/features/codespaces
35. https://colab.google/
36. https://github.com/settings/tokens/new
37. https://jupyterlab.readthedocs.io/en/stable/index.html
38. https://www.markdownguide.org/basic-syntax/
39. https://github.com/xbwei/Social-Data-Analytics-in-the-Cloud-with-AI/blob/main/COVID_19_Data_Analysis.ipynb
40. https://www.ecdc.europa.eu/en/publications-data/download-todays-data-geographic-distribution-covid-19-cases-worldwide
41. https://aws.amazon.com/s3/
42. https://pandas.pydata.org/pandas-docs/stable/reference/io.html
43. https://pandas.pydata.org/pandas-docs/stable/reference/frame.html
44. https://github.com/xbwei/Social-Data-Analytics-in-the-Cloud-with-AI/blob/main/COVID_19_Data_Analysis.ipynb
45. https://github.com/xbwei/Social-Data-Analytics-in-the-Cloud-with-AI/blob/main/COVID_19_Data_Analysis.ipynb
46. https://pandas.pydata.org/docs/user_guide/indexing.html
47. https://github.com/xbwei/Social-Data-Analytics-in-the-Cloud-with-AI

## References

Downey, Allen. 2015. *Think Python: How to Think Like a Computer Scientist*. 2nd edition. Beijing Boston Farnham Sebastopol Tokyo: O'Reilly Media.

GitHub, Inc. n.d. "GitHub.Com Help Documentation." GitHub Docs. Accessed June 29, 2024. https://docs.github.com/en.

Hunter, John, Darren Dale, Eric Firing, Michael Droettboom, and Matplotlib Development Team. n.d. "Tutorials — Matplotlib 3.9.0 Documentation." Accessed June 30, 2024. https://matplotlib.org/stable/tutorials/index.html.

Jupyter Team. n.d.-a. "Project Jupyter Documentation — Jupyter Documentation 4.1.1 Alpha Documentation." Accessed June 26, 2024. https://docs.jupyter.org/en/latest/.

Jupyter Team. n.d.-b. "User Documentation — Jupyter Notebook 7.2.1 Documentation." Accessed July 2, 2024. https://jupyter-notebook.readthedocs.io/en/stable/user-documentation.html.

Matthes, Eric. 2019. *Python Crash Course*. 2nd edition: A Hands-On, Project-Based Introduction to Programming. 2nd edition. San Francisco, CA: No Starch Press.

pandas. n.d. "User Guide — Pandas Documentation." Accessed June 30, 2024. https://pandas.pydata.org/pandas-docs/stable/user_guide/index.html.

Python Software Foundation. n.d. "Python Documentation." Accessed June 30, 2024. https://docs.python.org/3/.

# 4

# *Data Collection and Storage*

People generate vast amounts of data. According to the 2023 *Data Never Sleeps Report*,[1] an average person produces 102 MB of data in 60 seconds. People send 6,944 ChatGPT prompts, 360K Tweets, and 241M emails within one minute. We are in an era where every state of human activities can be measured, collected, and analyzed. Therefore, collecting and storing vast amounts of data is essential in data analytics. This chapter explores various data storage solutions in the cloud and techniques for collecting data with Application Programming Interfaces (APIs).

This chapter first introduces relational and NoSQL databases and how to set up database instances on AWS and MongoDB. We also learn to use graphic user interfaces (GUIs) to manage databases and APIs to collect data. GUIs are more manageable for beginners who have just started learning data analytics. Although we strive to utilize cloud-based resources, no cloud-based GUIs are available now, so we need to download and install the GUIs on our local computers.

In addition, we recommend students check the PostgreSQL Documentation (The PostgreSQL Global Development Group, n.d.)[2] and MongoDB Manual[3] (MongoDB, Inc., n.d.) to learn more about those two databases and two textbooks for users who wish to explore further data collocation techniques: *Data Wrangling with Python* (Kazil and Jarmul 2016) and *Mining the Social Web: Data Mining Facebook, Twitter, LinkedIn, Instagram, GitHub, and More* (Russell and Klassen 2019).

## Cloud-based Data Storage

The goal of databases is to facilitate data sharing among different users. They are designed to reduce data redundancy by eliminating unnecessary duplicates, maintaining data integrity and consistency by minimizing incorrect data or data operations, and supporting fast and flexible data retrievals. By organizing data efficiently, databases ensure that users can access accurate and up-to-date information, streamlining operations and decision-making processes across various applications.


DOI: 10.1201/9781003437611-4

Various data storages are designed to support different data structures. For example, data can be *structured*, *semi-structured*, or *unstructured*. Structured data is organized in a predefined format, typically rows and columns, and can be stored in a *relational database*, such as PostgreSQL,[4] or a *data warehouse*, such as Amazon Redshift.[5] Relational databases and data warehouses provide fast queries due to their simple data structure and are optimal storage for business intelligence (BI) analysis.

However, most data, such as images, videos, audio, and texts, is *unstructured*. Unstructured data can be stored in a data lake, such as Amazon S3,[6] which we used in the previous chapter. Data lakes can handle massive, heterogeneous data without predefined structures, making them ideal containers for training machine learning models and developing AI applications. This flexibility allows data scientists and developers to access and process large volumes of diverse data, facilitating advanced analytics, pattern recognition, and model training with minimal data preparation and transformation efforts.

Semi-structured data is between structured and unstructured data types. It is flexible in storing various values but still follows a specific structure, such as *XML* or *JSON* documents. For high flexibility and good performance, we can store semi-structured data in NoSQL databases, a relatively new database like MongoDB.[7]

Figure 4.1 summarizes how different data types can be stored in various data storages. Both relational databases and data warehouses can store structured data. However, relational databases store data by rows and provide good transactional latency. Data warehouses store data by columns and are efficient in conducting data aggregations. They usually require a cluster of computers to execute queries in parallel. Therefore, relational databases are widely adopted for online transaction processing (OLTP), suc as processing

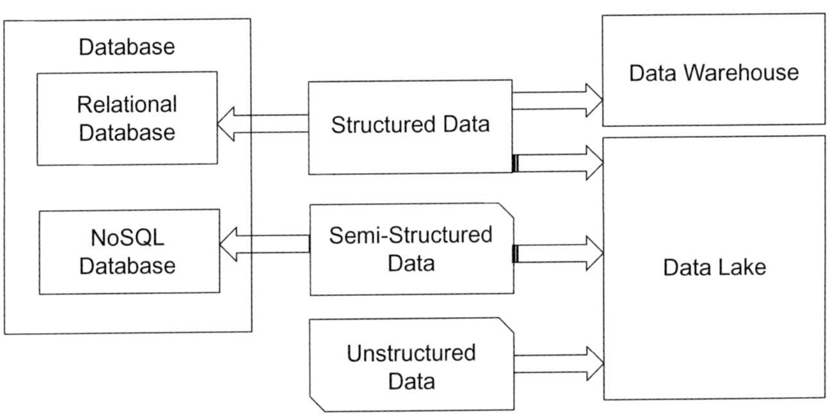


**FIGURE 4.1**
Types of Data and Data Storages

online orders or payments. Data warehouses are time- and resource-consuming to build but suitable for online analytical processing (OLAP), such as marketing analysis or profit forecasting. In addition, although data lakes can store all types of data, we prefer to store structured data in relational databases or data warehouses and store semi-structured data in NoSQL databases for better performance.

## Relational Database

Relational databases are *ACID* compliant, meaning all transactions are atomic, consistent, isolated, and durable. They organize data into two-dimensional tables where each *column* stores the same data type, and each *row* stores a record of the same entity. For example, in the Table 4.1 *People* table, each column stores specific information about people, such as *people ID, name,* and *city ID*. Each row records that information for a single person.

Every table has one (and only one) *primary key* to distinguish each row of the table. A primary key can include one or multiple columns, and the values in the primary key must be unique, not null, and atomically indexed. For example, the *People ID* is the primary key in the *People* table to distinguish people's records, and each person has a unique ID. The *City ID* is the primary key of the Table 4.2 *City* Table. However, if the *Zip Codes* are unique and not null for all cities, they can also serve as the primary key.

Multiple tables are related based on *foreign keys*. Foreign keys can be one or multiple columns, but the values can be null or duplicated. For example, the *City ID* on the *People* table is a foreign key, which refers to the primary key on

**TABLE 4.1**

People Table

| People ID/ Primary Key | First Name | Last Name | City ID/ Foreign Key |
|---|---|---|---|
| P1 | James | Brown | C1 |
| P2 | David | Johnson | C2 |
| P3 | Thomas | Brown | C1 |

**TABLE 4.2**

City Table

| City ID/ Primary Key | City Name | Zip Code |
|---|---|---|
| C1 | Chicago | 60007 |
| C2 | New York | 10005 |

the *City* table. There are also duplicated values in that foreign key. If we need to search for a person's corresponding city name, we need to *join* the People table and the City table based on the *City ID*, then identify the city names based on the person's ID.

Relational databases are *schema-on-write*, i.e., table structures must be carefully designed before a massive amount of data is inserted, and all the stored records must match the defined table structures. Relational database design follows a set of *normalization* rules. The first normalization form (NF) requires that each table has a primary key and no multiple values in a single cell. For example, Table 4.3 violates the first normalization form, where multiple values exist in single cells, and a primary key is not defined. Therefore, the values in the first row should be split into two rows, and a primary key should be added, such as the Table 4.1 People table.

The second normalization form aims to reduce data redundancy. Table 4.4 is an example of violating the second form, as the City Name and Zip Code values are redundant and do not depend on the primary key. To eliminate data redundancy, we should split the table and use foreign keys to relate both tables, such as the Table 4.1 People table and the Table 4.2 City table.

Third, fourth, and fifth normalization forms are "theoretically desirable, but not always practical." Please check this document[8] for more descriptions of those normalization rules.

The related tables can have three types of *relationships*: one-to-one, one-to-many, and many-to-many.

- In the one-to-one relationship, for each record in one table, there is only one corresponding record in another table. For example, one person can have one row in a social security table and one corresponding row in a driver's license table.

**TABLE 4.3**

A Table Violates the First NF

| People ID | First Name | Last Name | City ID |
|---|---|---|---|
| P1/ P3 | James/ Thomas | Brown | C1 |
| P2 | David | Johnson | C2 |

**TABLE 4.4**

A Table Violates the Second NF

| People ID/ Primary Key | First Name | Last Name | City ID | City Name | Zip Code |
|---|---|---|---|---|---|
| P1 | James | Brown | C1 | Chicago | 60007 |
| P2 | David | Johnson | C2 | New York | 10005 |
| P3 | Thomas | Brown | C1 | Chicago | 60007 |

- In the one-to-many relationship, multiple records in one table relate to only one record in another table. For example, one row in the City table relates to various rows in the People table.
- In the many-to-many relationship, one row in Table A relates to one or multiple rows in Table B, while one row in Table B can also relate to one or many rows in Table A. For example, one student may enroll in various courses and thus have multiple related records in a course table. One course has several students enrolled, which relates to different records in a student table.

The common practice for designing table structures in relational databases is using an *Entity Relationship (ER)* diagram. In ER diagrams, each table is called an *entity* linked to other entities with different *relationships*. Figure 4.2 shows how the People and City tables are organized on an ER diagram. ER diagrams can help design table structures at the *logical* level and export Structured Query Language (SQL) codes. Those SQL codes can be executed in a database to create those tables at the *physical* level.

Although relational databases are the most adopted data storage system, they have many limitations, such as a fixed data structure and *vertical scaling*. For example, to increase database storage and performance, we must replace the entire database instance with an instance with larger storage capacities and faster processors. Such upgradation is usually very expensive.

This textbook introduces the PostgreSQL database, the number 4 database, and the number 2 open-source database on the *DB-Engines* ranking.[9] A PostgreSQL server is a computer instance that runs the PostgreSQL services. We can create multiple databases on a single server and build numerous schemas in a single database. A *schema* is a logical container to store different tables, views, functions, or indexes. A table is the fundamental element to store structured data. We can define *constraints* or *triggers* to check or validate inserted values. PostgreSQL also supports spatial data with the PostGIS[10] extension. Please check the official PostgreSQL documents for more about the database objects.[11]

PostgreSQL can store integers, floats, strings, Boolean data, date and time, JSON documents, and spatial data. It can store integers and floats with

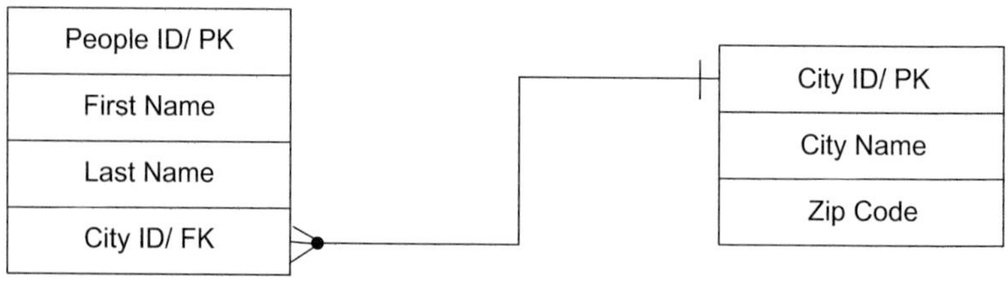


**FIGURE 4.2**
ER Diagram of the People Table and City Table

different sizes or precisions. We can also define string data with fixed or variable lengths. The PostgreSQL documents list all the supported data types.[12]

## Set up a PostgreSQL Instance on AWS

Utilizing database instances in the cloud can reduce costs and improve performance. *AWS RDS*[13] and MongoDB offer fully managed database solutions, so we don't need to purchase or maintain physical infrastructure on-premises or update operating systems and software patches. We can choose database engines we're familiar with and deploy instances in regions closer to our clients. We can also increase or reduce storage or the number of instances based on data size or client numbers in real time to optimize costs.

To set up a relational database instance, we need to start the AWS Learner Lab, open the AWS Console, and search Amazon RDS service. On the RDS service page, we can click *Create a database.* Amazon RDS (Amazon Web Services, Inc., n.d.) supports the most popular relational databases, including PostgreSQL, MySQL, MariaDB, Oracle, Microsoft SQL Server, and IBM Db2.[14] If users plan to use a commercial database, such as Microsoft SQL Server, they must purchase and bring their license to use those databases.

We can select *Easy Create* in the *Database creation method* to simplify the process and choose *PostgreSQL* for the *Engine type.* We can select the *Free tier* for the instance type to save credits. Please note: The free tier instance still costs our AWS credits. For *Credential management,* we choose *Self-managed* and manually type a master username and password. We will use the username and password later in the database GUI. We can leave everything else as default and click the *Create database* button.

While waiting for the instance creation, we can view the database *Summary* by clicking the instance identifier. In the *Connectivity & security* section, we need to change the *VPC security group* settings. VPC, which stands for *virtual private cloud,* is a logical section that separates our AWS services from other users' services.[15] A security group is like a firewall on our VPC to control the inbound and outbound traffic.[16] We must open the security group setting and add one *inbound rule* like Figure 4.3. This rule allows any IP address outside our VPC to access PostgreSQL instances. In production, we give the least privilege, such as providing the exact IP addresses to access our database. However, as students, we may not have a fixed IP address, so we allow *Any IP address* to connect to the database.

When the database is available, we can modify the instance by clicking the *Modify* button on the database *Summary* page. For example, we can change the master password or instance type, increase the storage size, and enable automated backups. AWS will deploy those changes without deleting our data.

**FIGURE 4.3**
Edit Security Group Setting for RDS

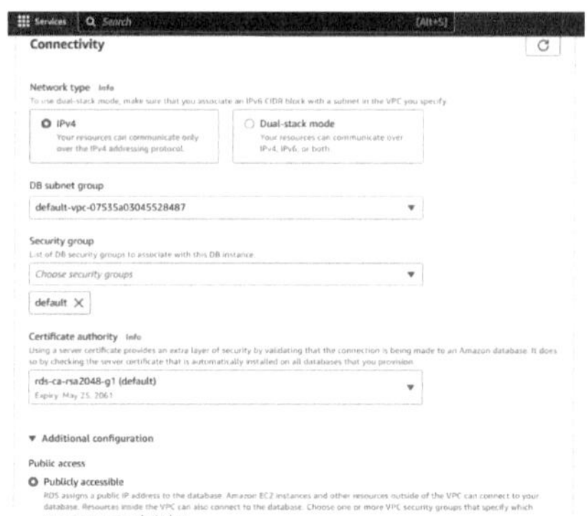


**FIGURE 4.4**
Grant Public Access to an RDS Instance

We must grant *Public* access to the database instance because we want to manage our database with pgAdmin, a popular PostgreSQL graphic user interface (GUI) outside our VPC. Figure 4.4 shows how to change the settings in the *Additional Configuration* under the *Connectivity section*. Please note: You should only grant public access when you *absolutely* need it.

We can download pgAdmin from the pgAdmin website,[17] choose the version suitable for our computer operating system, and install it. Once pgAdmin is installed, we can start pgAdmin and *Register* a server. In the pgAdmin *Connection* section, we must provide the host address (endpoint), port, master username, and password, as shown in Figure 4.5. The endpoint and port are available on the AWS RDS database *Summary page* under the *Connectivity & security* section.

After registering the PostgreSQL server, we see the server details in pgAdmin. As shown in Figure 4.6, we can monitor the instance status by checking the server sessions or the number of transactions. We can also explore the database, schemas, tables, and views on the server.

When the instance is not in use, we should *Stop* it manually from the *Actions* list on the RDS database *Summary* page. Please note: If you end the Learner Lab without stopping the database instance, the instance will continue

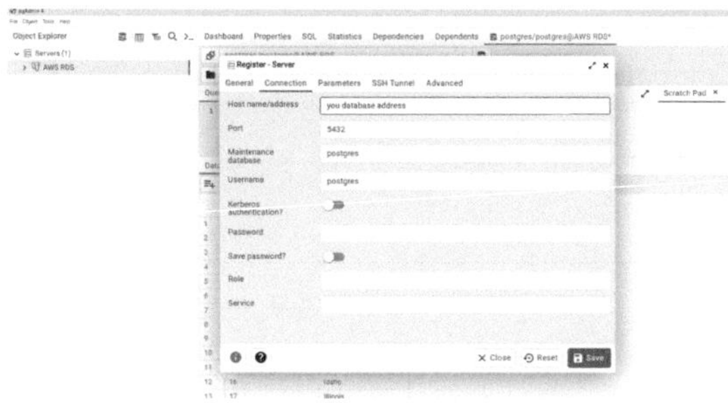


**FIGURE 4.5**
Connect to RDS Instances with pgAdmin

**FIGURE 4.6**
Monitor RDS Instances with pgAdmin

running and consuming the credits. Also, a stopped database instance will automatically restart after seven days.

## NoSQL Database

Unlike relational databases, NoSQL databases are *ACID-comprised* and *schema-on-read*. We can handle data in its 'native' format and store massive data with greater flexibility and higher scalability. The schema is then applied when we need to query the data. NoSQL databases can store semi-structured datasets. Depending on how the data is organized, they can be *key-value* stores like Amazon DynamoDB,[18] *graph* stores like Neo4j,[19] or document stores like MongoDB. Key-value stores can be used in applications requiring fast reads and writes. Graph stores can handle relationships between entities, such as social networks.

MongoDB stores data in JavaScript Object Notation (*JSON*) documents. A JSON document is like a *dictionary* in Python and is efficient in encoding and

**TABLE 4.5**

Term Comparison between Relational Databases and MongoDB

| Relational Database | MongoDB |
|---|---|
| Row | Document |
| Column | Field |
| Table | Collection |
| Database | Database |
| Index | Index |
| Join | Embedding |

decoding data for network transmission. Compared to a relation database, a single JSON *document* is like a row in a relational database table, and a *field* in a JSON document is like a column. Multiple documents are stored in a *collection*, just as many rows are stored in a table. However, in MongoDB, documents with different fields can be stored in the same collection, allowing more flexible data storage. In addition, information about various entities is *embedded* in a single document, so joining different collections in MongoDB is uncommon. Table 4.5 summarizes the similarities and differences in terminologies between relational databases and MongoDB.

In the following example, we store the same information in the Table 4.1 *People* table and Table 4.2 *City* table from a relational database in a MongoDB database. Instead of using two tables, we need one collection to store the people's information, and their city information is stored in the *City* field, a *nested document*. The documents can also have different structures. For example, the *P2* person doesn't have the *Last Name* information, while *P3* can have a *Gender* field.

```
{'People ID': 'P1',
'FirstNname': 'James',
'Last Name': 'Brown',
'City':{
    'City ID': 'C1',
    'City Name': 'Chicago',
'   Zip Code': 60007 }
},
{'People ID': 'P2',
'FirstNname': 'David',
'City':{
    'City ID': 'C2',
    'City Name': 'New York',
    'Zip Code': 10005 }
},
```

```
{'People ID': 'P3',
'FirstNname': 'Thomas',
'Last Name': 'Brown',
'Gender': 'Male',
'City':{
     'City ID': 'C1',
     'City Name': 'Chicago',
     'Zip Code': 60007}
}
```

MongoDB stores data in a *cluster*, or Replica Set, rather than a single instance. As shown in Figure 4.7, a MongoDB Replica Set includes one primary and multiple secondary instances. The primary instance receives and stores data from clients and constantly replicates the data to other secondary instances. The instances can be distributed in different geographic regions, so if the primary instance fails, one of the secondary instances can become the primary instance. We can query data from the primary instance directly or from the secondary instances to reduce the workload on the primary instance. Such cluster structure of MongoDB Replica Set enhances the *availability* of the data services.

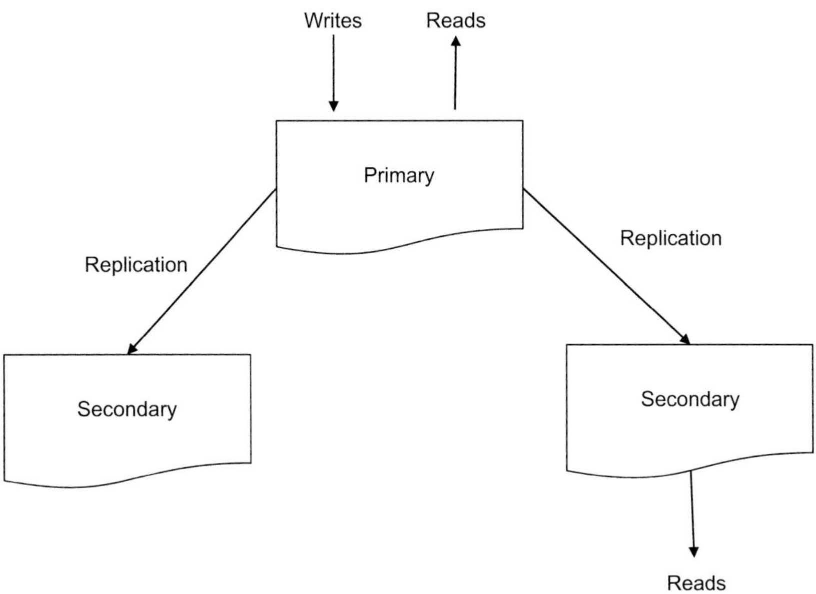


**FIGURE 4.7**
A MongoDB Replica Set

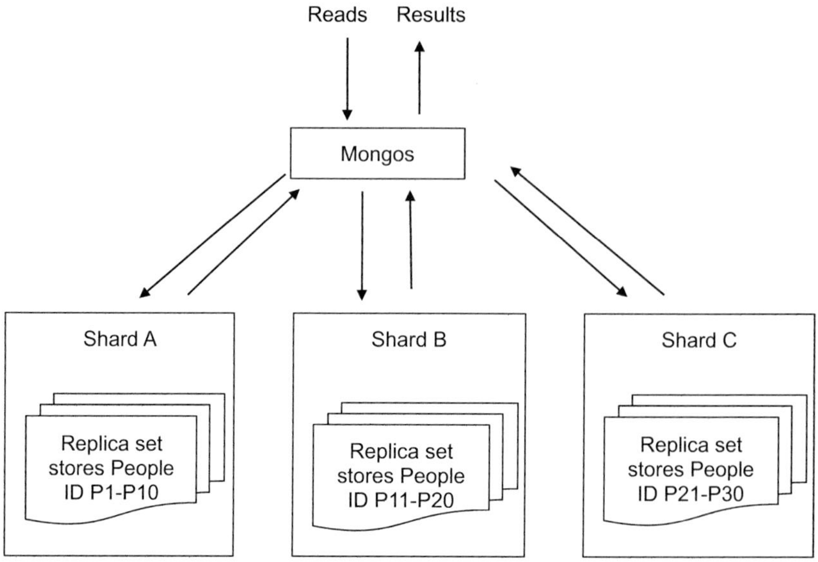


**FIGURE 4.8**
MongoDB Sharded Clusters

When the data size is beyond the storage capacity of a single MongoDB Replica Set, we can *scale horizontally* by adding more commodity servers, i.e., sharded clusters. Figure 4.8 shows the organization of sharded MongoDB clusters. The data is divided into multiple pieces and stored across many shards. Each shard is a MongoDB Replica Set that stores part of the data. For example, the Replica Set in Shard A can store people's information for ID from P1 to P10, Shard B stores ID from P11 to P20, and Shard C stores ID from P21 to P30. The Mongos instance receives the data inputs and queries and decides which shards to store or process. When Mongos gets the results from the shards, it aggregates and returns them to the clients. The cost of horizontal scale is lower than vertical scale because we can leverage commodity computing rather than purchasing supercomputers.

## Create a MongoDB Cluster

Chapter 2 introduced how educators can create organizations on MongoDB Atlas[20] and invite students to a specific organization. Once students accept the invitation and set up their accounts, they can create projects within the organization. Students can name their projects differently and manage only the projects they created.

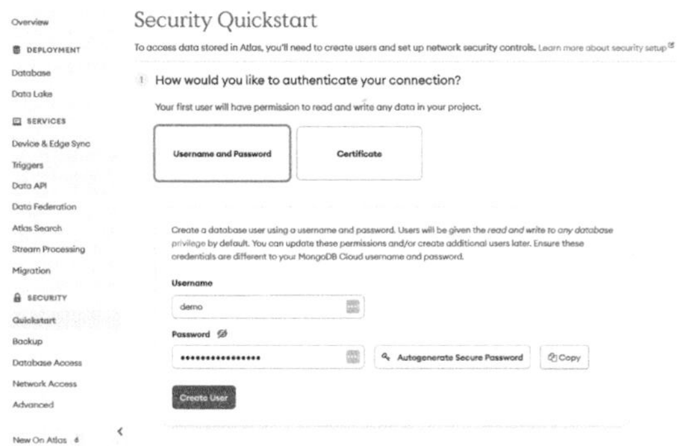


**FIGURE 4.9**
Set up MongoDB Master Username and Password

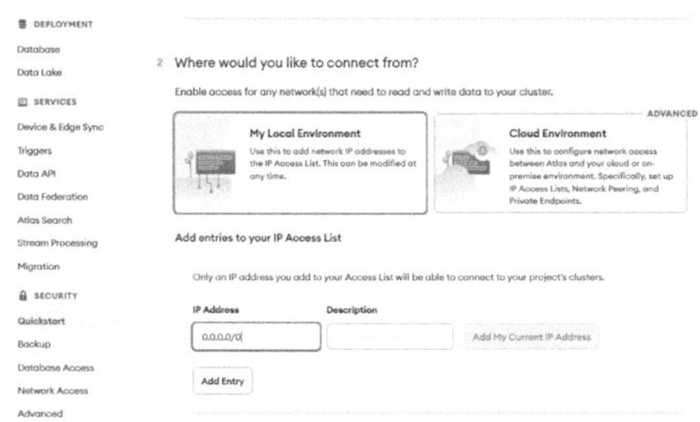


**FIGURE 4.10**
Edit MongoDB IP Access List

In a MongoDB project, we can create a cluster. We can choose the *M0* cluster, which has 512 MB storage and is "free forever."[21] Please note: The M0 cluster doesn't support sharded clusters and data backups. We can also select the cloud provider that hosts the cluster, such as AWS. As shown in Figure 4.9, we need to provide the master username and master password, the same as we did in AWS RDS.

Next, we need to define the IP Access List. Like the AWS RDS security group, this list controls which IP addresses can access MongoDB clusters. As shown in Figure 4.10, we added *0.0.0.0/0* to the IP list, allowing *any IP address* to access the cluster. We should only allow the exact IP addresses to access the cluster in production.

When the cluster is ready, we see three buttons on the database dashboard that allow us to *Connect* the cluster, *View Monitoring*, and *Browse Collections*. The Connect page provides the credentials needed to connect to the database. The Monitoring page displays the status of instances in the cluster. As shown in Figure 4.11, one primary and two secondary instances exist in the cluster. We can also open the *Collections* page to view the databases and collections on the cluster.

**FIGURE 4.11**
Status of MongoDB Cluster

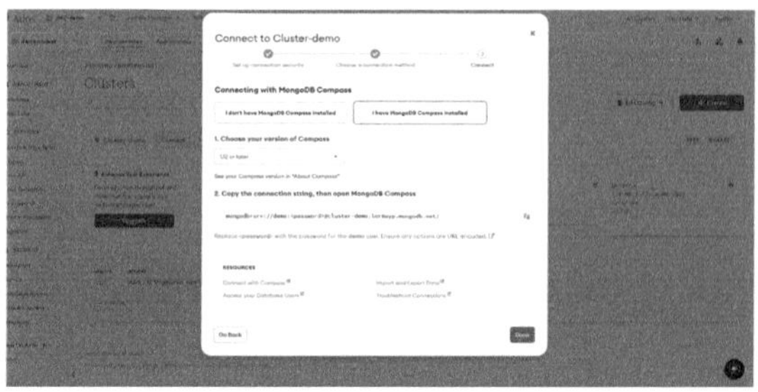


**FIGURE 4.12**
Copy MongoDB Connection String for Compass

**FIGURE 4.13**
Type Connection String in MongoDB Compass

We can now download and install the MongoDB GUI[22] named *Compass* on our computer. Then, we can visit the *Connect* page on the MongoDB website, choose Compass as the *connection method*, and *Copy the connection string*, as shown in Figure 4.12.

We can start MongoDB Compass from our computer, paste the connection string, and manually type the password within the string, as shown in Figure 4.13. We can then click the *Connect* button to access the databases and collections in the MongoDB Cluster. Because the M0 instance is free, we don't need to stop it manually. The instance will automatically pause after 60 days of inactivity and resume when we connect it.

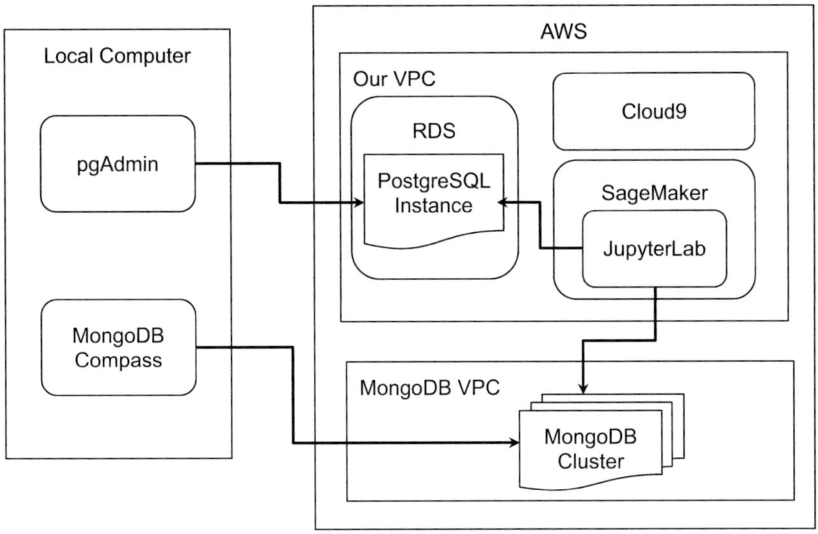


**FIGURE 4.14**
How Cloud Resources Connected with Our Local Computer

Now, we set up all the services in the cloud and installed the required tools on our local computer. As shown in Figure 4.14, we set up a PostgreSQL instance, a JupyterLab instance, and a Cloud9 instance in our AWS VPC. We also built a MongoDB cluster in another AWS VPC managed by MongoDB. We installed the GUIs for both databases on our local computer, namely the pgAdmin for PostgreSQL and the MongoDB Compass for MongoDB. We opened public access on the PostgreSQL instance and the MongoDB cluster so we can use the GUIs to access those databases. We can also log in to the AWS console and use JupyterLab to access both databases.

## Data Collection via API

There are generally three ways of collecting data from the Internet: downloading data directly from the source, web crawling, and collecting data with an Application Programming Interface (API).

Agencies or institutions may publicly share their data and provide dedicated web pages for users to download. *Data.gov*[23] is the first website we should search for open-source data. Other websites include the *Census Bureau,*[24] *Centers for Disease Control and Prevention (CDC),*[25] and *World Bank.*[26] Companies or communities also host datasets for public use, such as the Registry of Open Data on AWS[27] and Kaggle.[28] If you can find reliable data sources that provide exciting data with explicit downloading pages, downloading the data from their websites is the easiest way to collect data.

You can use Python or other programming languages to scrape webpages that contain interesting information and store the information in a database. This technique is called *web crawling*. Once you have identified a website that lists the information that interests you, you can use Python libraries like *urllib*[29] to download the web page and *beautifulsoup*[30] to extract the information. Web crawling does require an intermediate level of programming skills. The book, *Data Wrangling with Python* (Kazil and Jarmul 2016) provides detailed instructions.

However, not all websites allow people to crawl their information, and you must follow their guidelines. The common practice is to check the *robots.txt* file published by each website. For example, if you want to know whether X (previously known as Twitter)[31] allows web-crawling, you should check their guidelines at https://x.com/robots.txt.[32] The X robots.txt file only allows Google Search Engine Robot to crawl specific pages and explicitly disallows any web crawling from other agents or individuals. In the era of Generative AI, data has become extremely valuable in training foundation models or building AI applications. Companies thus become more cautious about their private datasets and profit by selling their data. For example, in 2023, both X[33] and Reddit[34] stopped free access to their datasets. Therefore, violating web-crawling guidelines may not only be an *ethical issue* but a *legal issue*.

More agencies, companies, or organizations are willing to share their data with an *Application Programming Interface*, or API for short. API allows data providers to select who they want to share data with and what data they want to share. For data consumers like us, API provides more flexibility in collecting the exact data in desired formats during specific periods. We can easily integrate the data collection into analytical pipelines to conduct real-time data analysis. Figure 4.15 illustrates the process of collecting data with API: We first need to apply for valid API tokens from the data providers. Some providers, such as X,[35] may charge a fee for collecting their data. Next, we can use Python or other programming languages to call the APIs with query parameters and API tokens. The data provider will validate the tokens and return the requested data in JSON format. We can then transform the received data into a desired format and store the data in our data containers.

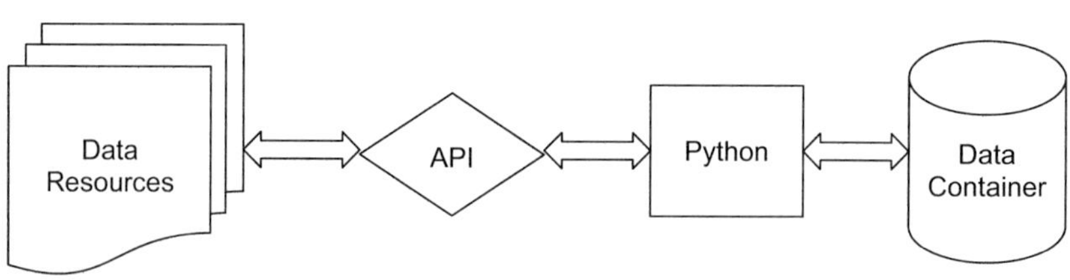


**FIGURE 4.15**
Data Collection with API

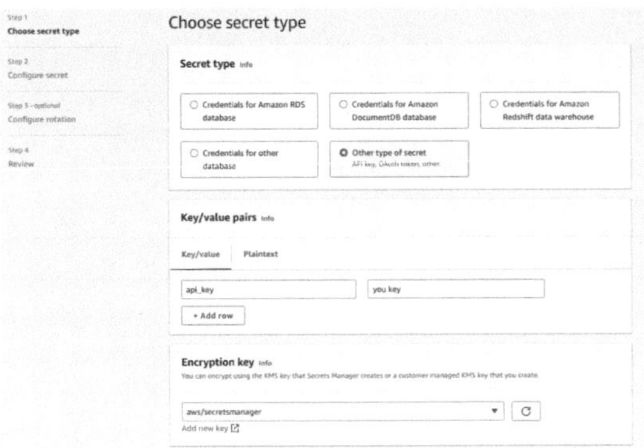


**FIGURE 4.16**
Store Census API Keys on AWS Secrets Manager

## Collect Census Data

The US Census Bureau provides free APIs[36] to download their publicly available datasets.[37] We can follow the example queries[38] to customize the API URLs or use the census[39] Python library, a simple wrapper, to request the data. We must also apply for an API key on the *Key Signup page*[40] to authorize our data request. The Census API key is a long string that should not be shared with others.

We should pass the API key and database credentials in Python without explicitly revealing them. The best practice is to store sensitive information with a secret management service, such as the AWS Secrets Manager,[41] and call the secret management service to retrieve credentials when needed. To store the credentials in AWS Secrets Manager, we can start the AWS Leaner Lab, open AWS Console, search and open the Secrets Manager service, and click *Store a new secret*. For the Census API key, we choose the *Other type of secret* in the *Secret Type* and paste the key into the key/value pairs. We can name the key *api_key*, as shown in Figure 4.16.

We can provide a name for this secret on the *Configure secret* page, such as *census*. A Python *Secret function* is provided on the *Review* page, as shown below. We can use this function in our notebook to retrieve the API keys and database credentials.

```
import boto3
from botocore.exceptions import ClientError

def get_secret():

    secret_name = "census"
```

```
region_name = "us-east-1"

# Create a Secrets Manager client
session = boto3.session.Session()
client = session.client(
      service_name='secretsmanager',
      region_name=region_name
)

try:
    get_secret_value_response = client.
    get_secret_value(
      SecretId=secret_name
    )
except ClientError as e:
  raise e
secret = get_secret_value_response['SecretString']
# Your code goes here.
```

Next, we can store the credentials of our relational database following the same steps. The only difference is that we need to choose Amazon RDS in the *Secret Type* and manually type the master user name and password in the *Credentials*. We also need to select the database instance on the *Database list*, as shown in Figure 4.17.

Because relational databases are *schema-on-write*, we must define the table structures before inserting data. There are many paid or free ER diagram editors, such as ERDLab.[42] This section uses ERDLab, a free cloud-based ER diagram editor. We can log into ERDLab with our GitHub account or register

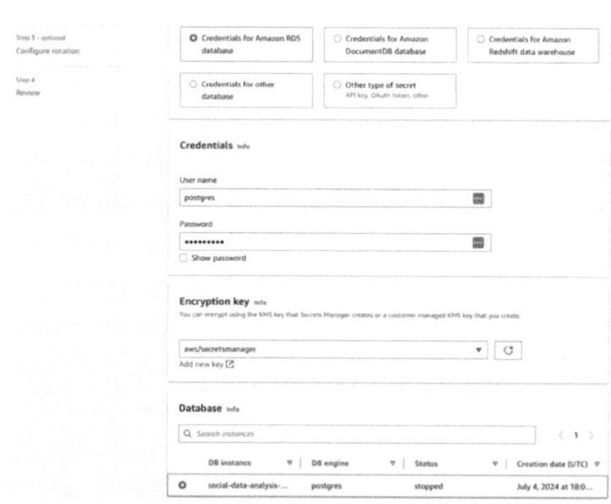


**FIGURE 4.17**
Store Database Credentials on AWS Secrets Manager

# 3D Organoids and Stem Cells in Regenerative Endodontics

This book provides an overview of the latest advancements in the use of three-dimensional (3D) organoids and stem cells for regenerative endodontics. It begins with an introduction to the foundational techniques for 3D organoid culture and explores cutting-edge developments in 3D bioprinting specifically tailored for dental applications. The book further examines the role of organoids in drug screening, toxicity testing, and the modulation of immune responses. It discusses the role of the microenvironment in organoid development, detailing how biophysical and biochemical factors, along with engineered niches, can influence the growth and functionality of organoids. The book reviews the use of various types of stem cells in regenerative endodontic therapies and explores the complex cellular and molecular interactions within organoid cultures, including stem cell-macrophage crosstalk and key signaling pathways involved in tissue regeneration. The role of the extracellular matrix (ECM) and different types of scaffolds, including nanofibrous, injectable scaffolds, bioinks, and decellularized ECMs, are discussed for their importance in supporting 3D organoid integration and biocompatibility.

**Key Features:**

- Explores cutting-edge techniques for 3D organoid culture and bioprinting designed for dental applications.
- Examines the role of organoids in drug screening, toxicity testing, and immune response modulation within endodontic research.
- Highlights the importance of the microenvironment in organoid development.
- Presents innovative scaffolds and biomaterials, such as nanofibrous scaffolds, bioinks, and decellularized ECMs, for enhanced integration and biocompatibility in 3D organoid-based therapies.
- Explores cutting-edge developments in 3D bioprinting technologies tailored for dental tissue engineering.

This book is intended for researchers of tissue engineering, regenerative medicine, and endodontics and clinicians.

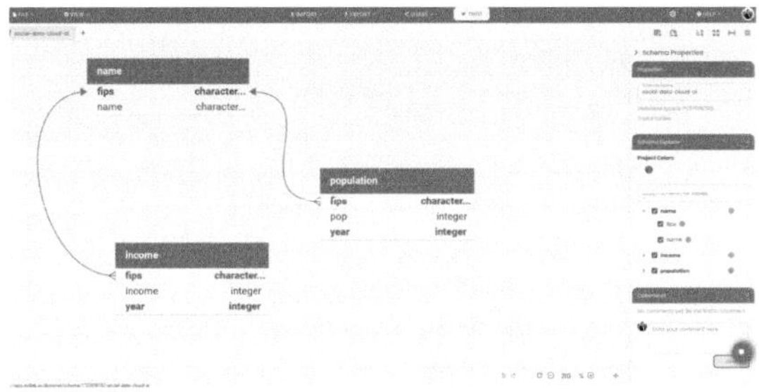


**FIGURE 4.18**
The ER Diagram Editor

for a new free account. After logging in, we can create a new project and select *PostgreSQL* as the database type.

Figure 4.18 displays the Editor interface with a created ER diagram. To add a table, we can click the *new table* icon and define column names, data types, and associated foreign keys in the *edit table* window. The current ER diagram in Figure 4.18 includes *income, population,* and *name* entities. The income and population tables store each state's median household income and total population, respectively. Because we need to store the income and population data for multiple years, the Federal Information Processing Standards (*FIPS*) and the *year* fields jointly serve as the primary keys on both tables. We use the *name* table to store state names separately to reduce data redundancy and set the FIPS as the primary key on the name table. Based on the *fips* field, we then relate the name table to the income and population table.

Once the ER diagram is completed, we can click the *Export* button and select *Export SQL*. In the pop-up window, we make sure *PostgreSQL* is the database type and click *Export* again. A SQL file will be downloaded to our local computer. Below are the SQL codes generated from the ER diagram. Based on the ER diagram, the codes create three tables and add two foreign keys. We can execute those SQL codes in Python to create the tables and define the relationships.

```
-- Create tables
CREATE TABLE IF NOT EXISTS name
(
    fips VARCHAR(4) NOT NULL UNIQUE,
    name VARCHAR(100) UNIQUE,
    PRIMARY KEY(fips)
);
CREATE TABLE IF NOT EXISTS population
(
    fips VARCHAR(4) NOT NULL,
    pop INTEGER NOT NULL,
```

```
    year INTEGER NOT NULL,
    PRIMARY KEY(fips, year)
);
CREATE TABLE IF NOT EXISTS income
(
    fips VARCHAR(4) NOT NULL,
    income INTEGER NOT NULL,
    year INTEGER NOT NULL,
    PRIMARY KEY(fips, year)
);
-- Create FKs
ALTER TABLE population
    ADD FOREIGN KEY (fips)
    REFERENCES name(fips)
    MATCH SIMPLE
;
ALTER TABLE income
    ADD    FOREIGN KEY (fips)
    REFERENCES name(fips)
    MATCH SIMPLE
;
```

We can (re)start the PostgreSQL instance on AWS RDS and the JupyterLab instance on AWS SageMaker. In JupyterLab, we can create a new notebook and name it *Collect_Census_Data*. In the notebook, we need to install the following Python packages:

- census[43]: a Census API wrapper to call Census APIs
- us[44]: a package providing US states' and territories' names and FIPS codes
- psycopg2[45]: PostgreSQL databases adapter to create tables, insert data, and execute queries

```
pip install census
pip install us
pip install psycopg2
```

Next, we modified the Python *Secret function* provided by Secret Manager and pasted it into the notebook to retrieve credentials. We can pass the *secret_name* in the *get_secret* function, and the corresponding credentials, such as database passwords or Census API Keys, will be returned.

```
import boto3
from botocore.exceptions import ClientError
import json
def get_secret(secret_name):
    region_name = "us-east-1"
    # Create a Secrets Manager client
    session = boto3.session.Session()
    client = session.client(
           service_name='secretsmanager',
           region_name=region_name
    )
    try:
        get_secret_value_response = client.
        get_secret_value(
           SecretId=secret_name
        )
    except ClientError as e:
      raise e
    secret = get_secret_value_response['SecretString']
    return json.loads(secret)
```

We can use the *psycopg2* package (Federico Di Gregorio, Daniele Varrazzo, and The Psycopg Team, n.d.) to connect to our database and create a *connection* object and a *cursor* object. The connection object manages the connection to the database, and the cursor object can execute SQL codes. Without revealing the credentials, we use the *get_secret* function to pass the required database credentials, such as the host URL, username, password, and engine type.

```
import psycopg2
conn = psycopg2.connect(host=get_secret('postgresql'
)['host'],
    user=get_secret('postgresql')['username'],
    password = get_secret('postgresql')['password'],
    dbname=get_secret('postgresql')['engine'])
cur = conn.cursor()
```

We can use the cursor object to *execute* the SQL codes generated from the ER diagram and use the connection object to *commit* the changes in the database. Because the SQL codes are very long, we define a variable named *sql* to pass the SQL statements to the cursor object.

```
sql = """
-- Create tables
CREATE TABLE IF NOT EXISTS name
(
    fips VARCHAR(4) NOT NULL UNIQUE,
    name VARCHAR(100) UNIQUE,
    PRIMARY KEY(fips)
);
CREATE TABLE IF NOT EXISTS population
(
    fips VARCHAR(4) NOT NULL,
    pop INTEGER NOT NULL,
    year INTEGER NOT NULL,
    PRIMARY KEY(fips, year)
);
CREATE TABLE IF NOT EXISTS income
(
    fips VARCHAR(4) NOT NULL,
    income INTEGER NOT NULL,
    year INTEGER NOT NULL,
    PRIMARY KEY(fips, year)
);
-- Create FKs
ALTER TABLE population
    ADD   FOREIGN KEY (fips)
    REFERENCES name(fips)
    MATCH SIMPLE
;
ALTER TABLE income
    ADD   FOREIGN KEY (fips)
    REFERENCES name(fips)
    MATCH SIMPLE
;
"""
cur.execute(sql)
conn.commit()
```

Finally, we can retrieve data with Census APIs (Bureau, n.d.) and insert the data into the created tables. The following codes authorize the Census API with the *get_secret* function.

```
from census import Census
from us import states
```

```
c = Census(get_secret('census')['api_key'])
```

Next, we insert the state names and FIPS codes into the *name* table. We use a *for* loop to iterate the *us* package's state names and FIPS codes and formulate an INSERT SQL code with *f string*. The cursor object then executes the INSERT SQL code. Because *D.C.* is not in the *States and Territories* list, we use a separate SQL code to insert the D.C. name and FIPS. We then use the connection object to commit those changes.

```
for state in states.STATES_AND_TERRITORIES:
    sql = f"""insert into name (name,fips)
              values ('{state}','{state.fips}') """
    cur.execute(sql)
sql = f"""insert into name (name,fips)
          values ('{states.DC.name}','{states.DC.f
          ips}') """
cur.execute(sql)
conn.commit()
```

We then use a for loop to iterate the population data (*Census table: B01003_001E*) for all states between 2005 and 2019 and insert the data into the *population* table.

```
for year in range(2005,2020):
    print('collecting data in year of: ',year)
    for data in c.acs1.get('B01003_001E',{'for':
    'state:*'},year=year):
      sql = f"""insert into population (fips,pop,year)
      values ('{data['state']}',{data['B01003_001E']}
      ,{year}) """
      # print(sql)
    cur.execute(sql)
conn.commit()
```

Similarly, we use the for loop to retrieve the median household income (*Census table: B19019_00E*) between 2005 and 2019 and insert the data into the *income* table.

```
for year in range(2005,2020):
    print('collecting data in year of: ',year)
```

```
for data in c.acs1.get('B19013_001E',{'for':
'state:*'},year=year):
     sql = f"""insert into income
     (fips,income,year)
        values ('{data['state']}',{data['B19013_00
        1E']},{year}) """
     # print(sql)
     cur.execute(sql)
conn.commit()
```

Please note: If errors occur when executing SQL codes, we need to use the *rollback* function to reverse all pending transactions and double-check the cause of the errors in the SQL codes. Please check the Psycopg online document for more information about the connection rollback.[46] When we finish the data collection, we should close the connection(s) to the database to free system resources.

```
cur.close()
conn.close()
```

### Collect Twitter Data

X (formerly called Twitter) API (X Corp, n.d.) was the most popular API for collecting social media data. People can register a developer account on the X Developer Platform[47] and apply for an API key. Since 2023, the Free API has no longer been able to collect Twitter data. According to Twitter API documents,[48] the *Basic* API now costs $100 monthly to collect up to 10,000 Tweets per month. The *Pro* API can collect Tweets in real-time, and the monthly rate cap is 1 million. The cost of the Pro API is $5,000 per month. Once a Twitter Developer account is registered and a Basic or Pro API has been purchased, we can create a *Bearer Token* on the *Developer's portal*[49] and store the token on AWS *Secrets Manager* like we stored the Census API key.

We also need a connection string to allow Python to communicate with MongoDB clusters. The connection string for our MongoDB cluster can be found on the database *Connect* page of the MongoDB website. Unlike connecting with Compass, we need to select the *Drivers* and then *Python* when we choose the connection method, as shown in Figure 4.19. We also need to store the MongoDB connection string on AWS *Secrets Manager.* Remember to manually type the password when storing the MongoDB connection string.

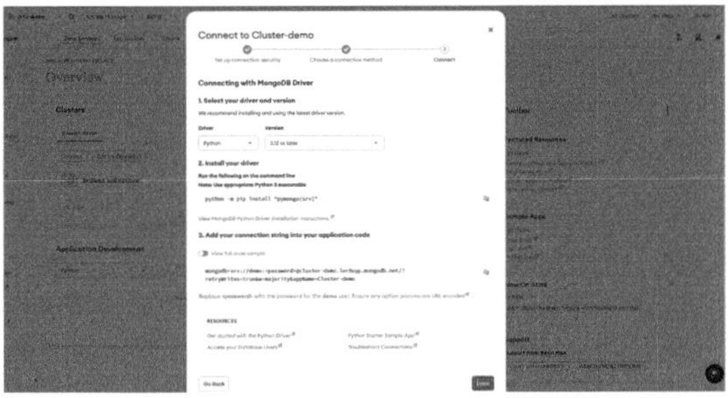


**FIGURE 4.19**
MongoDB Connection String for Python

Because the MongoDB database is *schema-on-read*, we don't need to define the collection structures. We can start the JupyterLab instance and collect Tweets directly. In JupyterLab, we can create a new notebook and name it *Collect_Twitter_Data*. In the notebook, we first need to install the *pymongo*[50] library to manage the MongoDB cluster and the *tweepy*[51] library to call the Twitter APIs.

```
pip install pymongo
pip install tweepy
```

Next, we paste the same *Secret Manager* function we used to collect the Census data. We can then import the Python libraries and credentials. In the following code, we use the *get_secret* function to pass the Twitter Bearer token to the *bearer_token* variable and the MongoDB connection string to the *mongodb_connect* variable.

```
import pymongo
from pymongo import MongoClient
import json
import tweepy

bearer_token = get_secret('twitter_api')['bearer_t
oken']
mongodb_connect = get_secret('mongodb')['connect
ion_string']
```

We are now ready to connect to the MongoDB cluster. The following code connects to our cluster and creates a database named *demo* and a collection named *tweet_collection*. If databases or collections with such names already exist, we insert data into them. An index is also added to the *tweet.id* key to guarantee that unique Tweets are inserted.

```
mongo_client = MongoClient(mongodb_connect)
db = mongo_client.demo # use or create a database
named demo
tweet_collection = db.tweet_collection #use or
create a collection named tweet_collection
tweet_collection.create_index([("tweet.id", pymongo.
ASCENDING)],unique = True) # make sure the collected
tweets are unique
```

We can now use the Twitter API to collect Tweets. We need to define the query, i.e., specify what Tweets interest us. The following query will collect the Tweets that contain the word *election*. Please check the official documents on how to define queries about searching Tweets.[52]

```
query = 'election' #query tweets contain the word
'election'
```

Next, we insert the collected Tweets into the MongoDB cluster. We can set a different *max_result* parameter to define how many Tweets will be returned in each request, but the maximum value is *100*. The *expansions* parameter defines additional data to collect, such as the authors' information and tagged locations. The *tweet_fields* and *user_fields* parameters specify which data fields to retrieve. Please check the official document on how to search Tweets[53]. The code then uses a *for* loop to iterate the retrieved Tweets and their authors' information, formats the data in JSON documents, and stores the JSON documents with a MongoDB INSERT command.

```
tweet_client = tweepy.Client(bearer_token)
tweets = tweet_client.search_recent_tweets(query=
query, max_results=100,
    expansions=['author_id'],
    tweet_fields = ['created_at','entities','lang
    ','public_metrics','geo'],
    user_fields = ['id', 'location','name',
'public_metrics','username'])
next_token = tweets.meta['next_token']
for user, tweet in zip(tweets.includes['users'],t
weets.data):
```

```
tweet_json = {}
tweet_json['tweet']= tweet.data
    tweet_json['user'] = user.data
    try:
          tweet_collection.insert_one(tweet_json)
          print(tweet_json['tweet']['created_at'])
    except:
          pass
```

We can try the following code to continue fetching more Tweets along the Tweets *Timeline*.[54] Please note: Depending on the API you purchased, you may reach your rate limit very soon, as the code can retrieve all the recent Tweets with the same query.

```
for i in range(5): # change the number here to fetch
more tweets
    tweets = tweet_client.search_recent_tweets(q
    uery=query, max_results=100,
                expansions=['author_id'],
                tweet_fields =
                ['created_at','entities','lang','pu
                blic_metrics','geo'],
                user_fields = ['id', 'location','name',
                'public_metrics','username'],
                next_token=next_token)
    next_token = tweets.meta['next_token']
    for user, tweet in zip(tweets.includes['users']
    ,tweets.data):
          tweet_json = {}
          tweet_json['tweet']= tweet.data
          tweet_json['user'] = user.data
          try:
            tweet_collection.insert_one(tweet_json)
            print(tweet_json['tweet']['created_at'])
          except:
            pass
```

We also need to close the connection to the MongoDB cluster after completing data collection:

```
mongo_client.close()
```

We can upload the notebooks to the GitHub repository and check the collected data with pgAdmin or Compass. We should also promptly stop the JupyterLab and PostgreSQL instances to save our credits. Users can check the demo notebooks published on the textbook's GitHub repository[55] and refer to the Census Data API User Guide(Bureau, n.d.) or Twitter API documents (X Corp, n.d.) for detailed API usage.

This chapter introduces various data storage solutions, teaches how to build a PostgreSQL database instance and a MongoDB cluster in MongoDB Atlas, and demonstrates how to collect Census and Twitter data with APIs. In Chapter 5, we learn how to query the collected data with SQL commands and MongoDB query language.

## Notes

1. https://www.domo.com/learn/infographic/data-never-sleeps-11
2. https://www.postgresql.org/docs/
3. https://www.mongodb.com/docs/v7.0/core/document/
4. https://www.postgresql.org/
5. https://aws.amazon.com/redshift/
6. https://aws.amazon.com/s3/
7. https://www.mongodb.com/
8. https://learn.microsoft.com/en-us/office/troubleshoot/access/database-normalization-description
9. https://db-engines.com/en/ranking
10. https://postgis.net/
11. https://www.postgresql.org/docs/current/tutorial.html
12. https://www.postgresql.org/docs/current/datatype.html
13. https://aws.amazon.com/rds/
14. https://docs.aws.amazon.com/rds/
15. https://docs.aws.amazon.com/vpc/latest/userguide/what-is-amazon-vpc.html
16. https://docs.aws.amazon.com/vpc/latest/userguide/vpc-security-groups.html
17. https://www.pgadmin.org/download/
18. https://aws.amazon.com/dynamodb
19. https://neo4j.com/
20. https://www.mongodb.com/
21. https://www.mongodb.com/docs/atlas/reference/free-shared-limitations/
22. https://www.mongodb.com/products/tools/compass
23. https://data.gov/
24. https://data.census.gov/
25. https://data.cdc.gov/
26. https://data.worldbank.org/
27. https://registry.opendata.aws/

28. https://www.kaggle.com/
29. https://docs.python.org/3/library/urllib.html
30. https://pypi.org/project/beautifulsoup4/
31. https://x.com/
32. https://x.com/robots.txt
33. https://x.com/XDevelopers/status/1621026986784337922?lang=en
34. https://www.reddit.com/r/reddit/comments/145bram/addressing_the_community_about_changes_to_our_api/
35. https://developer.x.com/en/products/twitter-api
36. https://www.census.gov/data/developers/data-sets.html
37. https://api.census.gov/data.html
38. https://www.census.gov/data/developers/guidance/api-user-guide.Example_API_Queries.html
39. https://pypi.org/project/census/
40. https://api.census.gov/data/key_signup.html
41. https://aws.amazon.com/secrets-manager/
42. https://app.erdlab.io
43. https://pypi.org/project/census/
44. https://pypi.org/project/us/
45. https://pypi.org/project/psycopg2/
46. https://www.psycopg.org/docs/connection.html
47. https://developer.twitter.com
48. https://developer.x.com/en/products/twitter-api
49. https://developer.x.com/en/portal/dashboard
50. https://pypi.org/project/pymongo
51. https://www.tweepy.org
52. https://developer.x.com/en/docs/twitter-api/tweets/search/integrate/build-a-query
53. https://developer.x.com/en/docs/twitter-api/tweets/search/quick-start/recent-search
54. https://developer.x.com/en/docs/twitter-api/tweets/timelines/quick-start/reverse-chron-quick-start
55. https://github.com/xbwei/Social-Data-Analytics-in-the-Cloud-with-AI

## References

Amazon Web Services, Inc. n.d. "Amazon RDS and Aurora Documentation." Accessed July 3, 2024. https://docs.aws.amazon.com/rds/.

Bureau, US Census. n.d. "Census Data API User Guide." Census.Gov. Accessed July 3, 2024. https://www.census.gov/data/developers/guidance/api-user-guide.html.

Federico Di Gregorio, Daniele Varrazzo, and The Psycopg Team. n.d. "Psycopg – PostgreSQL Database Adapter for Python — Psycopg 2.9.9 Documentation." Accessed July 7, 2024. https://www.psycopg.org/docs/.

Kazil, Jacqueline, and Katharine Jarmul. 2016. *Data Wrangling with Python*. 1st edition. Beijing Boston Farnham Sebastopol Tokyo: Oreilly & Associates Inc.
MongoDB, Inc. n.d. "What Is MongoDB? - MongoDB Manual v7.0." Accessed July 3, 2024. https://www.mongodb.com/docs/manual/.
Russell, Matthew, and Mikhail Klassen. 2019. *Mining the Social Web: Data Mining Facebook, Twitter, LinkedIn, Instagram, GitHub, and More*. 3rd edition. Beijing Boston Farnham Sebastopol Tokyo: O'Reilly Media.
The PostgreSQL Global Development Group. n.d. "PostgreSQL: Documentation." Accessed July 3, 2024. https://www.postgresql.org/docs/.
X Corp. n.d. "Twitter Developer Tutorials." Accessed July 3, 2024. https://developer.x.com/en/docs/tutorials.

# 5

# *Data Process and Query*

The collected raw data is not always suitable for direct analysis, as they may be stored in separate data containers that require data consolidation, or individual records need to be aggregated to identify relationships and patterns. Therefore, we need to process the collected raw data before consuming it. This process often involves data *extraction*, *transformation*, and *load*, or *ETL*. Figure 5.1 shows an example of the ETL process.

- Data extraction refers to extracting a subset of information from the raw data and saving it separately to enhance the query performance or to process the data that can't be performed in the raw data container.
- Data transformation includes data cleansing to remove errors, joining data from multiple resources to enrich information, splitting data to reduce data redundancy, or calculating new values by applying business logic.
- When raw data is aggregated, cleaned, or transformed, it will be loaded into a specific data container depending on how we want to consume it. Generally speaking, if we use business intelligence (BI) tools to visualize the data, e.g., creating dashboards, we prefer to store the data in a data warehouse for better query performance. Suppose we plan to train machine learning models or develop AI applications, such as facial recognition or language translation. In that case, we store the data in a data lake for high flexibility and scalability.

No matter which data databases we use, we always want to build *indexes* before we query the data. An index is an added data structure that can speed up database query performance. Without an index, the database must scan all the records to retrieve the results for a single query. If appropriate indexes exist, the database can retrieve the results by traversing the index. PostgreSQL and MongoDB support the most common indexes. In addition, we can create *Spatial Indexes*[1] on PostgreSQL to query large spatial datasets and build *Text Indexes*[2] on MongoDB clusters to search words or phrases. We can also utilize large language models to create *vector embeddings*[3] and perform semantic searches using MongoDB Atlas *Vector Search*.[4] Please note: We

DOI: 10.1201/9781003437611-5

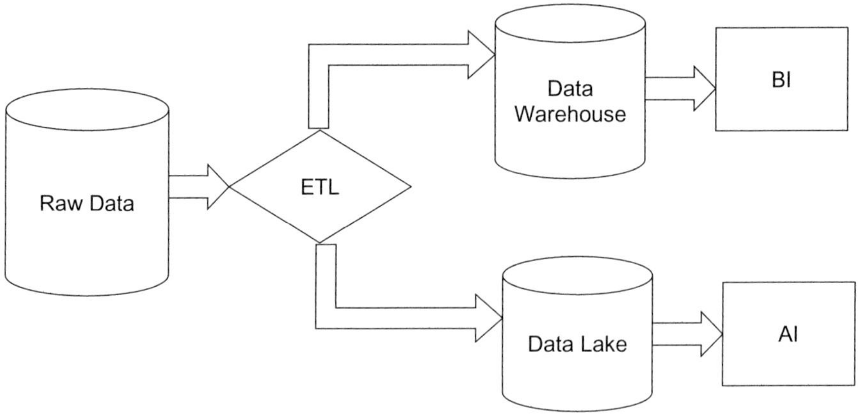


**FIGURE 5.1**
An Example of the ETL Process

should define indexes based on our query patterns, as unnecessary indexes can slow down data transactions.

This chapter introduces the basic techniques for processing and querying data in PostgreSQL and MongoDB. We recommend two textbooks, *MongoDB: The Definitive Guide* (Bradshaw, Brazil, and Chodorow 2019) and *Practical SQL: A Beginner's Guide to Storytelling with Data* (DeBarros 2018), for people who want to explore advanced data processes and query techniques. The *PostgreSQL Tutorial(PostgreSQLTutorial, n.d.), Official Document (The PostgreSQL Global Development Group, n.d.)*[5] and *MongoDB Manual*[6](MongoDB, Inc., n.d.) provide simple examples and quick references for learning PostgreSQL and MongoDB. This chapter executes queries in GUIs because of their friendly user interfaces and leverages Generative AI to build complicated queries.

## Query and Process Structured Social Data

Structured Query Language, or *SQL*, is a standard query language to manage relational databases. We can execute SQL codes in a database GUI, such as pgAdmin, or a programming language like Python. We can CREATE databases, users, tables, and indexes with SQL. We can also modify tables like INSERT, UPDATE, and DELETE records. The previous chapter used an ER diagram to generate SQL codes for creating tables and foreign keys. We then executed the SQL codes in Python and used a simple INSERT command to store the received Census data. Users can read this PostgreSQL Tutorial[7] to learn more about modifying data in SQL. This section focuses on querying data and joining tables in relational databases. If your PostgreSQL instance is stopped, you can start the instance and connect your database with pgAdmin.

## Query and Filter Data

Once the pgAdmin connects to the PostgreSQL instance, we can open the *Query Tool* from the pgAdmin Tools bar and type SQL commands in the *Query* window. The results are reported in the *Data Output* window. For example, we can view all the records in the *name* table by typing the following SQL statement:

```
select * from name
```

In the above example, we use the SELECT statement to indicate the columns we want to retrieve and the FROM statement to specify the table name. If we query the data of all columns, we can use an *asterisk* instead of typing the individual column names. Figure 5.2 shows the Data Output window displaying all the records of the *name* table.

We can apply simple calculations on the values in columns with the SELECT statement. For example, the following code divides the income values in the *income* table by 1,000 and reports the results in a new column named *income_in_thousand*:

```
select income/1000 as income_in_thousand from income
```

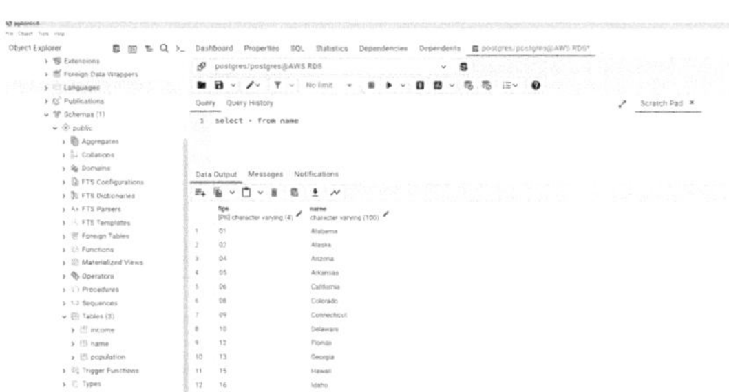

**FIGURE 5.2**
pgAdmin Query Tool

The following table shows the first five rows of the query results:

| income_in_thousand |
|---|
| 36 |
| 56 |
| 44 |
| 34 |
| 53 |

Please note: The calculated results of a SELECT statement are temporary. If we want to save the results, we should use the UPDATE statement instead. The ORDER BY operator can sort the retrieved data in descending (DESC) or ascending (ASC) orders, and the LIMIT operator can return a subset of the results. For example, the following SQL code sorts the population in the *population* table in descending order and retrieves the top 5 rows. It also reports the associated FIPS codes. Please note: We can write an SQL statement into multiple lines for better readability. To execute multiple SQL statements, we can use semicolons to separate those SQL statements.

```
select fips, pop from population
order by pop desc
limit 5
```

The result is printed below:

| fips | pop |
|---|---|
| 6 | 39557045 |
| 6 | 39536653 |
| 6 | 39512223 |
| 6 | 39250017 |
| 6 | 39144818 |

The WHERE clause can filter data by rows and is frequently used with the SELECT, UPDATE, and DELETE statements. For example, we can find the FIPS of *Virginia* from the *name* table by using the following SQL code:

```
select fips from name
where name = 'Virginia'
```

The output is:

| fips |
|---|
| 51 |

Like in Python, string values are declared in quotation markers, but numbers are not. For example, if we want to find the FIPS where the income is greater than 8,000, we can use the following code:

```
select fips, income from income
where income > 80000
order by income desc
limit 5
```

The query output is:

| fips | income |
|---|---|
| 11 | 92266 |
| 24 | 86738 |
| 25 | 85843 |
| 34 | 85751 |
| 11 | 85203 |

However, to find the incomes where the FIPS is 51, we must add quotation marks around 51 because the FIPS code is a string.

```
select fips, income from income
where fips = '51'
order by income desc
limit 5
```

Query output:

| fips | income |
|---|---|
| 51 | 76456 |
| 51 | 72577 |
| 51 | 71535 |
| 51 | 68114 |
| 51 | 66262 |

We can also combine multiple conditions in one WHERE clause. The following SQL code filters Virginia's income data (FIPS='51') after 2010 and sorts the results by year.

```
select income, year from income
where fips = '51' and year >2010
order by year desc
limit 5
```

Query output:

| income | year |
|---|---|
| 76456 | 2019 |
| 72577 | 2018 |
| 71535 | 2017 |
| 68114 | 2016 |
| 66262 | 2015 |

The WHERE clause can include the BETWEEN, IN, LIKE, and IS operators. The LIKE operator filters data that matches a defined pattern. We can use a *percentage sign* to match any sequence of characters and an *underscore* to match a single character. For example, if we want to find the state name starts with "M," we can use the following SQL code:

```
select * from name
where name like 'M%'
```

Query output:

| fips | name |
|---|---|
| 23 | Maine |
| 24 | Maryland |
| 25 | Massachusetts |
| 26 | Michigan |
| 27 | Minnesota |
| 28 | Mississippi |
| 29 | Missouri |
| 30 | Montana |

The BETWEEN operator filters data in *a specific range*, and the IN operator filters data from *a list of values*. For example, if we want to check the population of Virginia (FIPS = '51') and Maryland (FIPS = '24') between 2010 and 2025, we can use the following SQL code:

```
select * from population
where
fips in ('24','51')
and year between 2010 and 2015
order by fips, year
```

Query output:

| fips | pop | year |
|---|---|---|
| 24 | 5785982 | 2010 |
| 24 | 5828289 | 2011 |
| 24 | 5884563 | 2012 |
| 24 | 5928814 | 2013 |
| 24 | 5976407 | 2014 |
| 24 | 6006401 | 2015 |
| 51 | 8024617 | 2010 |
| 51 | 8096604 | 2011 |
| 51 | 8185867 | 2012 |
| 51 | 8260405 | 2013 |
| 51 | 8326289 | 2014 |
| 51 | 8382993 | 2015 |

The above code sorts the results by two columns: FIPS and year. Because Maryland's FIPS code is smaller, the output displays Maryland's population records first. Maryland's population records are then sorted in ascending order based on the year values.

The IS operator is typically used to filter *null* values. The following code returns all the income table rows where the FIPS value is null. The output is an empty table because the income table doesn't have null values.

```
select * from income
where fips is null
```

## Join and Group Data

In addition to querying the data by individual rows, we can use aggregate functions to perform calculations on a set of rows. The standard aggregate functions include:

- COUNT: return the number of values
- AVG: return the average value
- MAX: return the maximum value
- MIN: return the minimal value
- SUM: return the sum of values

For example, we can use this SQL code to count the number of rows containing Virginia (FIPS = '51') records in the *population* table:

```
select count (fips) from population
where fips = '51'
```

Query output:

| count |
|---|
| 15 |

We can also build a *subquery* to find the row(s) containing the maximum income. The inner SELECT statement is executed first in the following codes to identify the maximal income value. The maximal value is then passed to the WHERE clause of the outer SELECT statement to return all the rows that contain such a value.

```
select * from income
where income = (select max(income) from income )
```

Query output:

| fips | income | year |
|---|---|---|
| 11 | 92266 | 2019 |

The GROUP BY clause divides rows into groups and applies aggregations to each group. For example, we can identify the maximal income in each state:

```
select fips, max(income) as max _ income from income
group by fips
order by max _ income desc
limit 5
```

Query output:

| fips | max_income |
|---|---|
| 11 | 92266 |
| 24 | 86738 |
| 25 | 85843 |
| 34 | 85751 |
| 15 | 83102 |

We can also use the HAVING clause to filter results by groups, just like using the WHERE clause to filter values by rows. The following SQL code calculates the total population per year and reports the result after 2010:

```
select year,sum(pop) from population
group by year
having year > 2010
order by year
limit 5
```

Query output:

| year | sum |
|---|---|
| 2011 | 315298609 |
| 2012 | 317581124 |
| 2013 | 319743925 |
| 2014 | 322405453 |
| 2015 | 324893003 |

As we introduced in the previous chapter, information in relational databases is *split* into separate tables to reduce data *redundancy*. Those tables are related to each other based on foreign keys. For example, when we collect the state income and population data, we store the state names in the name table and the population/income data in the population/income tables. We then relate the population/income tables with the name table using the foreign keys, i.e., the fips. If we want to enrich our results by showing the state name and their associated income, we need to JOIN the tables based on those foreign keys.

There are four types of joins, and Figure 5.3 illustrates those types of joins:

- *Inner join*: return the rows that match both tables
- *Left join*: return all the rows from the left table and the matched rows from the right table
- *Right join*: return all the rows from the right table and the matched rows from the right table
- *Full join*: return all rows from both tables, including the non-matching rows

The following SQL code joins the *name* table and the *income* table to show the state name and their income in different years:

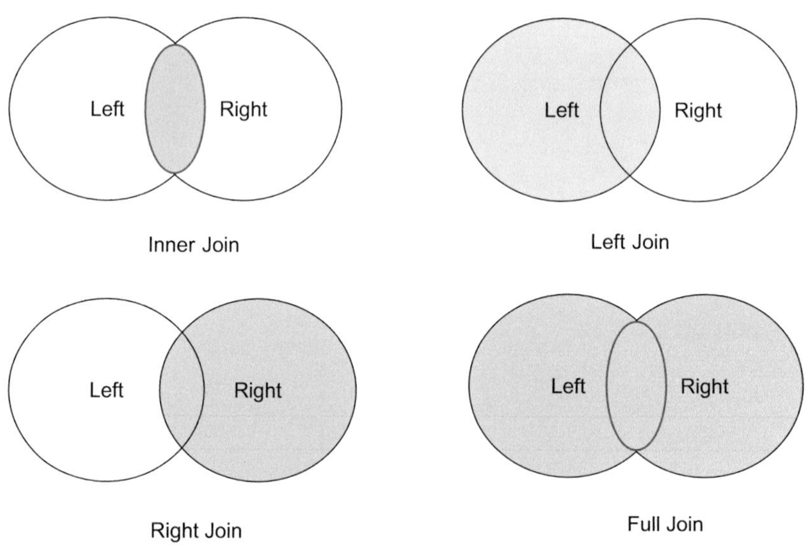


**FIGURE 5.3**
Four Types of Joins in SQL

```
select name.fips, name, income, year
from income
inner join name
on name.fips = income.fips
limit 5
```

Query output:

| fips | name | income | year |
|---|---|---|---|
| 1 | Alabama | 36879 | 2005 |
| 2 | Alaska | 56234 | 2005 |
| 4 | Arizona | 44282 | 2005 |
| 5 | Arkansas | 34999 | 2005 |
| 6 | California | 53629 | 2005 |

Please note: When joining multiple tables, we use *dot notation* to distinguish the columns with the same name but belonging to different tables. For example, the *name* table and *income* table have a column named *fips*. Therefore, *name.fips* refers to the fips column in the name table and *income.fips* refers to the fips column in the income table.

If there are queries we use frequently, such as joining the *name* table and the *income* table, we can store them as *views*. We can treat views as virtual tables not physically stored in the database. For example, to store the above joining query, we can create a view named *name_income* like the following. This code generates no output; instead, it makes a name_income view in the database.

```
create view name_income as
select name.fips, name, income, year from income
inner join name
on name.fips = income.fips
```

We can then query the created view like a table. The following example calculates the maximal income for each state with an aggregation function and reports the maximum incomes and associated state names.

```
select name, max(income) as max_income from
name_income
group by name
order by max_income desc
limit 5
```

Query output:

| name | max_income |
|---|---|
| District of Columbia | 92266 |
| Maryland | 86738 |
| Massachusetts | 85843 |
| New Jersey | 85751 |
| Hawaii | 83102 |

Views are great ways to handle complicated joins, aggregations, or filters, as they don't take much storage and are independent of the physical tables. We can also use views to control data access, such as hiding sensitive information or limiting subsets of data to be queried. For advanced SQL techniques, such as *set operations*, *subqueries*, and *conditional expressions*, please read the PostgreSQL Tutorial.[8]

## SQL Query with ChatGPT

SQL has been widely used in data analytics for decades. Therefore, large language models (LLMs) are well-trained with SQL statements. We can ask a chatbot powered by LLMs to write SQL codes. This section demonstrates using ChatGPT, developed by OpenAI,[9] to create SQL codes based on human prompts.

We can register a free account on the ChatGPT website[10] and tell the chatbot that we want to create a table to store state names and another to store state populations in different years. This section uses the *GPT-4o*[11] model released on May 13, 2024. The following is an example prompt:

```
I need to collect states' names and fips in one table
and states' populations in different years in another
table. How to define the table structure in SQL?
```

ChatGPT provides the following SQL code to store state names:

```
CREATE TABLE states (
    state_id SERIAL PRIMARY KEY,
    state_name VARCHAR(100) NOT NULL,
    fips_code CHAR(2) NOT NULL UNIQUE
);
```

It also explains the columns in the table:

```
state_id: A unique identifier for each state.
state_name: The name of the state.
fips_code: The FIPS code of the state.
```

The following codes are created to store the state population:

```
CREATE TABLE populations (
    population _ id SERIAL PRIMARY KEY,
    state _ id INT NOT NULL,
    year INT NOT NULL,
    population INT NOT NULL,
    FOREIGN KEY (state _ id) REFERENCES states(state _ id)
);
```

Here is the explanation of the population table provided by ChatGPT:

```
population_id: A unique identifier for each
population record.
state_id: A foreign key that references the state_id
in the states table.
year: The year of the population record.
population: The population of the state in the given
year.
```

ChatGPT also provides examples of usage, such as inserting data into those tables. The created SQL codes work fine. However, the *state_id* and *population_id* are unnecessary, as we could use the fips and the year columns as the primary keys. We can specify the requirements in the prompt to improve the SQL codes or upload the ER diagram we created in the previous chapter, as shown in Figure 5.4.

ChatGPT can recognize the tables and relationships in the ER diagram and create SQL codes to create those tables, which are the same codes generated by the *ERDLab* online ER diagram editor. We can then ask ChatGPT to write complicated SQL queries, such as:

```
Join the income table and the name table
```

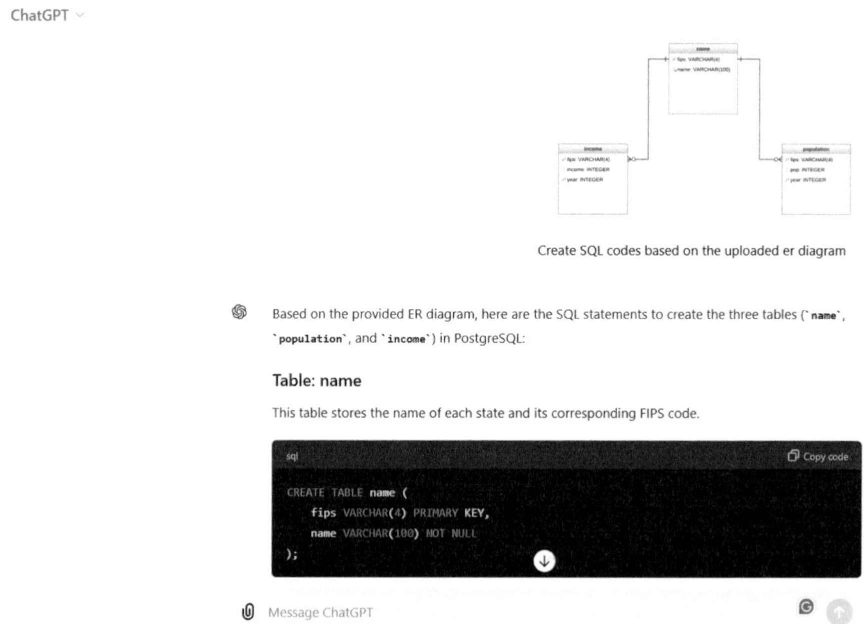


**FIGURE 5.4**
Create SQL Codes on ChatGPT by Uploading the ER Diagram

Below is the JOIN statement created by the chatbot:

```
SELECT
    name.fips,
    name.name,
    income.income,
    income.year
FROM
    name
INNER JOIN
    income
ON
    name.fips = income.fips;
```

It provides detailed explanations and dummy outputs to show the join query results. Once ChatGPT "understands" the structure of our table, it writes pretty accurate SQL queries. We can now focus on the questions we want to explore and leverage Generative AI to complete the sophisticated SQL codes. For example, we can ask ChatGPT the following question:

```
Which state has the highest income in recent years?
```

ChatGPT provides complicated SQL codes as follows:

```
WITH RecentIncome AS (
    SELECT
            fips,
            income,
            year,
            RANK()  OVER  (ORDER  BY  year  DESC)  AS
            year _ rank
FROM
            income
)
SELECT
    n.name,
    r.income,
    r.year
FROM
    RecentIncome r
JOIN
    name n ON n.fips = r.fips
WHERE
    r.year _ rank = 1
ORDER BY
    r.income DESC
    LIMIT 1;
```

We can execute the codes in pgAdmin and receive the following output:

| **name** | **income** | **year** |
|---|---|---|
| District of Columbia | 92266 | 2019 |

We can also ask a different question, such as:

```
Calculate the year-over-year population growth rate
for the VA state in the recent 5 years
```

ChatGPT assumes the FIPS code of Virginia is 51 and generates very sophisticated SQL codes:

```
WITH PopulationData AS (
    SELECT
          fips,
          pop,
          year,
LAG(pop) OVER (PARTITION BY fips ORDER BY year) AS
prev_pop
    FROM
          population
    WHERE
          fips = '51'
    ORDER BY
          year DESC
    LIMIT 6
)
SELECT
    year,
    pop,
    prev_pop,
    CASE
          WHEN prev_pop IS NOT NULL THEN ((pop::DECIMAL
          / prev_pop::DECIMAL) - 1) * 100
          ELSE NULL
    END AS growth_rate
FROM
    PopulationData
ORDER BY
    year DESC;
```

The output of the codes is accurate:

| year | pop | prev_pop | growth_rate |
|---|---|---|---|
| 2019 | 8535519 | 8517685 | 0.2093761391739700 |
| 2018 | 8517685 | 8470020 | 0.5627495566716500 |
| 2017 | 8470020 | 8411808 | 0.6920272074683600 |
| 2016 | 8411808 | 8382993 | 0.3437316481118400 |
| 2015 | 8382993 | 8326289 | 0.6810236829396600 |
| 2014 | 8326289 | 8260405 | 0.7975880117258200 |

With Generative AI, querying data in SQL is straightforward when accurate table structures are provided. Without revealing the data, we can obtain sophisticated SQL queries, execute them directly in databases, and receive correct results.

## Query and Process NoSQL Social Data

NoSQL databases like MongoDB support flexible data structures to store various datasets without a predefined schema. Although inserting data into NoSQL databases is easier, querying data from NoSQL databases can be challenging compared to relational databases. Like SQL, MongoDB has *CRUD* operations to CREATE, READ, UPDATE, and DELETE data. For example, the previous chapter used the CRUD operations in Python to CREATE a new database and a new collection and INSERT multiple Tweets into the database. We recommend the MongoDB Manual to readers who want to explore CRUD operators[12] further.

This section focuses on the READ operator and Aggregation pipelines in querying and processing the collected Twitter data. We highly recommend using Compass, the MongoDB GUI, to query the data for several reasons: The graphic user interface can highlight document structures that help build queries; we can export the queries from Compass into Python codes or other programming languages; we can leverage Generative AI to query MongoDB data in natural language.[13]

Figure 5.5 shows the MongoDB Compass interface once it connects to the *demo* database. With Compass, we can perform *Document Queries*, create *Aggregation Pipelines*, analyze *Schema*, manage *Indexes*, and create *Validation* rules.

In the *demo* database, there is one collection named *tweet_collection*. The collection contains 792 Tweets. We can open the Schema panel and *Analyze* the structure of the Tweet documents. All Tweet documents contain a *_id* field, a *tweet* field, and a *user* field. Please refer to the X developer documents for more about the Twitter data dictionary.[14]

- The _id field is the Object ID of MongoDB documents. The values are unique, automatically indexed, and recorded when a document is created. It is like a primary key in a relational database.

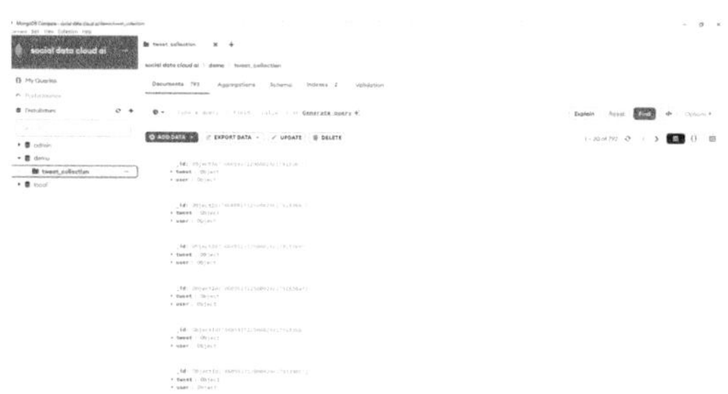

**FIGURE 5.5**
MongoDB Compass Interface

**FIGURE 5.6**
Create a Text Index on MongoDB

- The tweet field is a nested document that contains a Tweet Object,[15] including Tweet *ID*, when the tweet is *created_at*, Tweet *text*, *author_id*, *entities*, *public_metrics*, etc. The entities field is also a nested document, including information about *hashtags*, *mentioned users*, *URLs*, etc. The public_metrics fields record the number of times a Tweet has been *retweeted*, *replied* to, *liked*, or *quoted*.
- The user field stores a Twitter User Object,[16] including User *ID*, *name*, when the account is *created_at*, the *location* they specified in their profile, *public_metrics*, etc. The public_metrics field lists a user's *following* and *follower* counts.

Before we start the query, we need to create a *TEXT* index[17] to enable text search. We can open the Index panel and click *Create Index*. Two indexes have already been created: The *_id* index is automatically built on the _id field and the *tweet.id* index was built when we collected Tweets in Chapter 4. As shown in Figure 5.6, we can manually create a text index by selecting the *tweet.text* as the index file and *text* as the index type.

## Document Query

Document queries are like the SELECT statement that can filter JSON documents based on conditions. The basic syntax is:

```
{<field name>: <conditions>}
```

For example, if we want to find the Tweets written in *English*, we can use the following query:

```
{"tweet.lang":"en"}
```

**FIGURE 5.7**
Compass Query Bar

In the above example, we query the *lang* field, a child field of the *tweet* field. We use a dot notation to specify a child field, like *tweet.lang*. The condition is that the filtered value equals *en*, which stands for English. As shown in Figure 5.7, we can type the query in the *Documents Query Bar* and click the *Find* button. There are 726 documents returned, meaning 726 of the 792 Tweets are written in English.

In addition to the equality condition, MongoDB supports the following query operators, and all the operators start with a *dollar* symbol. Please check the MongoDB document for a complete list of query operators.[18]

- $eq: equals to a specified value
- $gt: greater than a specified value
- $lt: less than a specific value
- $in: equals a value in a particular *list*
- $ne: not equal to a specific value
- $exist: documents have the specific *field*
- $and: logical operators, return the documents match both conditions
- $or: logical operators, return the documents match either condition
- $text: performs a text search on the field, and a text index is required
- $size: returns any lists with the specified number of items

The syntax to use a query operator is:

```
{<field name>: {<$operator>:<value>}}
```

For example, if we want to search the Tweets that received more than one favorite count, we can use the *$gt* operator on the *like_count* field in the *tweet.public_metrics* field.

```
{ "tweet.public_metrics.like_count" : {$gt:1} }
```

This query returns only one Tweet that has been liked twice. The Tweet document is printed below; some fields are removed to save page space.

```
{
    " _ id": {
    "$oid": "6605931729d0024c1791863b"
    },
    "tweet": {
    ...
        "public _ metrics": {
        "retweet _ count": 0,
        "reply _ count": 1,
        "like _ count": 2,
        "quote _ count": 0,
        "bookmark _ count": 0,
        "impression _ count": 117
    },
...
}
```

If we want to find geotagged Tweets, we can search Tweets where the *geo*[19] field is not null with the *$ne* operator:

```
{"tweet.geo":{$ne:null}}
```

The query returns two geotagged Tweets as follows:

```
{
  "_id": {
     "$oid": "660592e529d0024c179184a1"
},
  "tweet": {
...
     "geo": {
     "place_id": "dc3747428fa88cab"
     },
  ...
     }
  }
}
{
     "_id": {
     "$oid": "660592e729d0024c179184e8"
  },
```

```
  "tweet": {
...
      "geo": {
         "place_id": "e52e231315b975c1"
      },
...
      }
  }
}
```

We can also combine multiple queries with a logical operator, and the syntax is:

```
{$logical_operator: [ {query1} , {query2} ]}
```

Suppose we want to identify Twitter users with a valid location field and more than 500,000 followers. In that case, we can use the *$ne* operator to find users whose *location* is not null and the *$gt* operator to identify the users with more than 500,000 followers. We then put both conditions in a list within the *$and* operator, as shown in the following example:

```
{$and :[ {"user.location":{$ne:null}} , {"user.pub
lic_metrics.followers_count":{$gt:500000}}]}
```

The above query returns one Twitter user:

```
{
     "_id": {
     "$oid": "6605931829d0024c1791867d"
  },
...
  "user": {
...
     "public_metrics": {
        "followers_count": 22111487,
        "following_count": 29,
        "tweet_count": 881925,
        "listed_count": 5586,
        "like_count": 1482
     },
...
  }
}
```

MongoDB supports TEXT search to query words or phrases on fields with a text index. Each collection can have only one text index. The syntax of the text search is:

```
{ $text: {$search: <string>} }
```

We already created a text index on the tweet.text field, so we can search Tweets that contain the word "ia":

```
{ $text: {$search: 'ia'} }
```

One Tweet contains the word *IA*, and the text is written in Hindi:

```
{
    "_id": {
    "$oid": "660592e329d0024c1791845d"
 },
 "tweet": {
...
"text": "RT @Damodar56192318: #election_commission
_permission_do\nराजसथान सरकार बेरोजगारों का साथ दो junior
Accountant और IA के रजिल्ट की अनुमति दि...",
...
    "lang": "hi",
...
 }
}
```

For fields containing lists, we can use the *$eq* operator to search a single item against specific values or use the *$size* operator to query lists based on the number of items. For example, this query identifies the Tweets that have four hashtags:

```
{"tweet.entities.hashtags":{$size:4}}
```

Only one Tweet contains four hashtags:

```
{
    "_id": {
    "$oid": "660592ee29d0024c179185b9"
 },
    "tweet": {
```

```
...
       "hashtags": [
       {
         "start": 224,
         "end": 239,
         "tag": "KalyanBanerjee"
       },
       {
         "start": 240,
         "end": 253,
         "tag": "TMCGagsMedia"
       },
       {
         "start": 254,
         "end": 266,
         "tag": "TMCKillsFoE"
       },
       {
         "start": 267,
         "end": 278,
         "tag": "SayNotoTMC"
    }
 ],
...
 }
}
```

In addition to filters, the Query Bar provides additional tools like *Project*, *Sort*, and *Limit*:

- Project: Set which fields to return, 1 to include, and 0 to exclude a field.
- Sort: Sort the documents based on the values in a field, 1 for ascending and -1 for descending.
- Limit: Set the upper limit of the returned document number.

In the example of Figure 5.8, we use the *Project* to display only the *text* field and the *like_count* field. We also *Sort* the returned Tweets in descending order based on the *like_count* and *Limit* the results to the top one Tweet.

MongoDB supports advanced queries, such as Geospatial queries,[20] Atlas Search,[21] and Atlas Vector Search.[22] Please check the MongoDB Manual[23] to learn more.

**FIGURE 5.8**
Compass Query Bar with Other Options

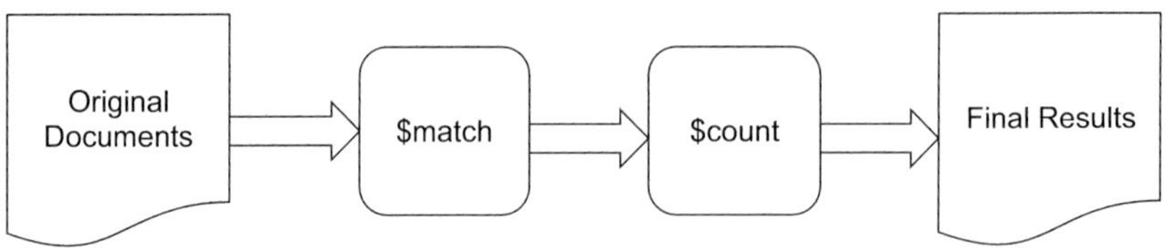


**FIGURE 5.9**
An Illustration of the MongoDB Aggregation Pipeline

## Aggregation Pipeline

Aggregation pipelines transform documents into aggregated results, like the Aggregate functions and the Group By operator in SQL. They are called pipelines because we modify documents in separate stages, and the previous stage's output is the following stage's input. For example, as shown in Figure 5.9, we can start a $match stage to filter the Tweets containing a specific keyword and then use a $count stage to calculate the number of Tweets containing this keyword.

In Compass, we can visually design the pipeline. We can open the Aggregations panel and *Add a Stage*. From the dropdown list, we can select the *$match* operator and type a text search in the stage to extract Tweets containing the word *election*:

```
{
$text: {
     $search: "election"
}
}
```

The outputs of this stage are all the Tweets containing the word *election*. We then add a new stage, select the *$count* operator, and provide a field name for the operator, such as:

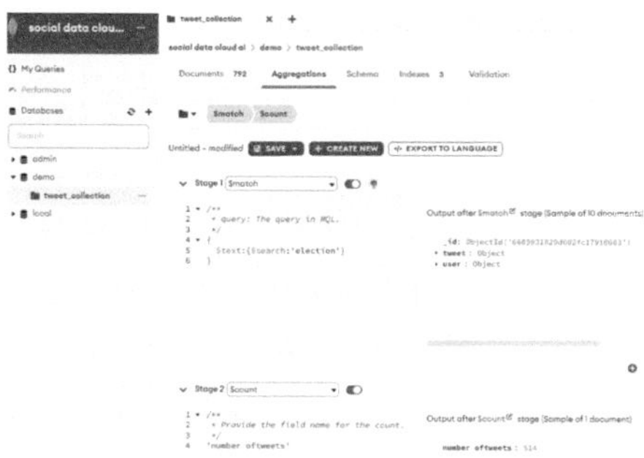

**FIGURE 5.10**
MongoDB Aggregation Pipeline Example

```
"number of tweets"
```

The $count operator returns the number of documents from the previous stage. Figure 5.10 displays the stages and results, showing that 514 Tweets contain the word *election*.

If we click the *TEXT* button on the Aggregation panel, we can obtain the query codes of this pipeline as follows:

```
[
 {
     $match:
       /**
       * query: The query in MQL.
       */
     {
        $text: {
        $search: "election"
        }
     }
 },
{
     $count:
        /**
        * Provide the field name for the count.
        */
        "number of tweets"
 }
]
```

The standard aggregation operators include:

- $match: filter documents and pass the matched document to the next stage
- $project: pass the documents with requested fields to the next stage
- $sort: sort all input documents and return them to the next stage
- $limit: limit the number of documents passed to the next stage
- $count: count the number of documents in the input stage and pass it to the next stage
- $sortByCount: group incoming documents based on specified expressions and compute the count of documents in each group
- $unwind: unwind a list from the input document to create a document for each item in the list
- $group: group input documents based on specified expressions and pass those groups to the next stage

For example, we can $unwind *hashtags* in all Tweets and use the $sortByCount and $limit operators to identify the most popular hashtag:

```
[
  {
   $unwind:
     /**
       * path: Path to the array field.
       * includeArrayIndex: Optional name for index.
       * preserveNullAndEmptyArrays: Optional
       * toggle to unwind null and empty values.
       */
     {
       path: "$tweet.entities.hashtags"
     }
     },
     {
       $sortByCount:
         /**
     * expression: Grouping expression or string.
     */
     "$tweet.entities.hashtags.tag"
   },
   {
       $limit:
       /**
```

```
        * Provide the number of documents to limit.
      */
      1
  }
  ]
```

The most popular hashtag is:

```
  {
  "_id": "election_commission_permission_do",
  "count": 8
  }
```

To find who sent the most Tweets, we can $group all Twitter users and count the number of Tweet documents in each group. We then use the $sort and $limit operators to report the top one Twitter user:

```
  [
    {
      $group:
        /**
          * _id: The id of the group.
          * fieldN: The first field name.
        */
      {
          _id: "$user.name",
          user_count: {
          $sum: 1
        }
      }
    },
    {
      $sort:
        /**
          * Provide any number of field/order pairs.
          */
        {
          user_count: -1
        }
      },
      {
        $limit:
```

```
            /**
            * Provide the number of documents to limit.
            */
        1
      }
]
```

The user who sent the most number of Tweets is:

```
{
"_id": "Rahul jain",
"user_count": 7
}
```

Please check the MongoDB Manual for a complete list of the aggregation operators,[24] pipeline optimizations,[25] and advanced topics, such as creating views[26] or processing streaming data.[27]

## Query Data in Natural Language

MongoDB queries are complex, and the syntax is trivial. For example, we must add the dollar symbol before the field name in aggregation pipelines but not in document queries. We also need curly brackets in the $match stage and $unwind stage. However, using curly brackets in the $count stage or $sortByCount stage gives us errors.

Fortunately, we can now ask questions in natural language. Advances in natural language processing and AI have enabled us to interact with computers and systems using everyday language, making it easier to retrieve information, get insights, and perform tasks without needing to know complex query languages or technical details. This development significantly enhances user accessibility and convenience in leveraging technology. Generative AI can convert the prompts into document queries or aggregation pipelines.[28] We can click the *Generate query* link on the Documents panel to enable the natural language query. We must log in with our free MongoDB account in the next window. After that, we can type our questions in the *Generate Query bar*. For example, we can type the following prompt and press *Generate*:

```
find the document containing 'election' with a text
search and sort the results based on the like count.
```

**FIGURE 5.11**
Document Query with Natural Language

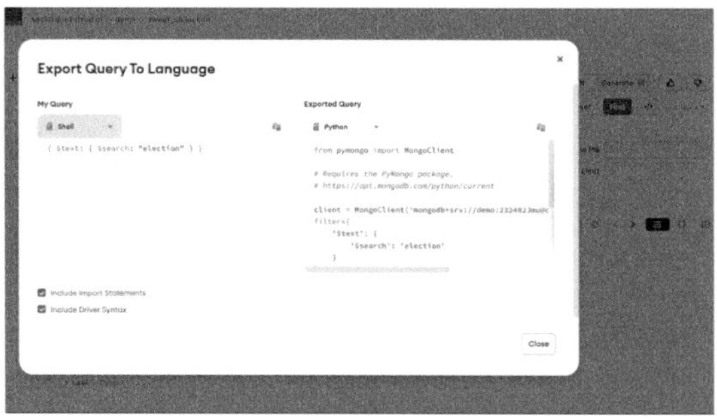


**FIGURE 5.12**
Export Query to Python

As shown in Figure 5.11, AI generates a quite accurate query with a text search operator and a sort operator. It also recognizes that the *like_count* field is located in the *tweet.public_metrics* field.

Clicking the *Export to Language* button allows us to export the query into Python or other programming languages. Figure 5.12 shows the Export Query window. The MongoDB connection string is also provided if we check the *Include Import Statements* box.

However, without specific instructions, the AI assistant may not utilize the appropriate features to generate an optimized query. For example, if we type the following prompt that doesn't require using the text index:

```
find the document containing 'election' in tweet
text and sort the results based on the like count.
```

The generated query utilizes a regular expression search[29] that is much slower than a text search.

```
{
    "tweet.text": {
      $regex: "election",
      $options: "i"
    }
}
```

On the other hand, AI understands our questions well and can identify the fields relevant to the questions. For example, on the Aggregation panel, we ask the following prompt that doesn't indicate which field to use:

```
what time have most tweets been posted
```

The generated query searches the *created_at* field and extracts the *hour* informant to *group* and *sort*.

```
[
  {
    "$project": {
      "hour": { "$hour": { "$toDate": "$tweet.
      created_at" } }
    }
  },
  {
    "$group": {
      "_id": "$hour",
      "count": { "$sum": 1 }
    }
  },
  {
    "$sort": { "count": -1 }
  },
  {
    "$limit": 1
  }
]
```

It then returns the correct result that all the Tweets were posted at 3 pm.

```
{
    "_id": 15,
    "count": 792
}
```

We can try an even more vague question, such as:

```
find the most popular Twitter user
```

The AI-generated query, as shown below, is very impressive. Since we didn't specify how to quantify *popularity* in the prompt, the AI assistant uses *followers_count* as the evaluating criteria and sorts *usernames* based on it.

```
[
  {
    "$group": {
      "_id": "$user.username",
      "totalFollowers": { "$sum": "$user.public_met-
      rics.followers_count" }
    }
  },
  {
    "$sort": { "totalFollowers": -1 }
  },
  {
    "$limit": 1
  }
]
```

The output is also accurate:

```
{
  "_id": "aajtak",
  "totalFollowers": 22111487
}
```

If we change the prompt to be more specific:

```
find the Twitter user who sent the most number of
Tweets
```

The AI assistant generates almost the exact query we did in the previous section. The only difference is that we returned the *user.name* field and the AI assistant returned the *user.username* field.

```
[
  {
    $group: {
      _id: "$user.username",
```

```
      tweet_count: { $sum: 1 }
    }
  },
  {$sort: { tweet_count: -1 } },
  {$limit: 1 }
]
```

The query also returns the correct result:

```
{
    "_id": "Rahuljain026",
    "tweet_count": 7
}
```

We also test whether AI can handle fields containing lists. For example, we want to find the most popular hashtag by using the following prompt:

```
find the most popular hashtag
```

Unfortunately, we often need to try multiple times to get a correct response. Sometimes, the AI assistant misinterprets the input and thus generates incorrect answers. This trial-and-error process can be frustrating and time-consuming, highlighting the limitations and imperfections of current AI technology. Despite these challenges, continued advancements are improving the accuracy and reliability of AI responses. Below is one of the correct responses:

```
[
    {
      "$unwind": "$tweet.entities.hashtags"
    },
    {
      "$group": {
      "_id": "$tweet.entities.hashtags.tag",
      "count": { "$sum": 1 }
    }
  },
  {
    "$sort": { "count": -1 }
  },
```

```
    {
      "$limit": 1
    }
  ]
```

The output is:

```
  {
      "_id": "election_commission_permission_do",
      "count": 8
  }
```

Most AI-generated queries work smoothly. However, it is essential to examine the outputs with common sense and consult official tutorials when in doubt. This critical approach ensures the reliability and accuracy of the results, allowing users to identify and correct any potential errors or inconsistencies in the AI-generated responses. We also need a basic understanding of MongoDB query techniques and frequently check the *Explain Results*[30] to achieve the best query performance.

In the previous section, when we used ChatGPT to write SQL queries, we only described table structures or uploaded ER diagrams without providing data to the AI models. In MongoDB Compass, our data is processed by Microsoft Azure OpenAI services.[31] Therefore, users should check the Azure OpenAI Data Privacy[32] before using private datasets with the natural language query function.

## Notes

1. https://postgis.net/workshops/postgis-intro/indexing.html
2. https://www.mongodb.com/docs/manual/core/indexes/index-types/index-text/
3. https://platform.openai.com/docs/guides/embeddings
4. https://www.mongodb.com/products/platform/atlas-vector-search
5. https://www.postgresqltutorial.com/
6. https://www.mongodb.com/docs/manual/
7. https://www.postgresqltutorial.com/
8. https://www.postgresqltutorial.com/
9. https://openai.com/
10. https://openai.com/chatgpt/
11. https://openai.com/index/hello-gpt-4o/

12. https://www.mongodb.com/docs/manual/tutorial/query-documents/
13. https://www.mongodb.com/docs/compass/current/query-with-natural-language/enable-natural-language-querying/
14. https://developer.x.com/en/docs/twitter-api/data-dictionary/introduction
15. https://developer.x.com/en/docs/twitter-api/data-dictionary/object-model/tweet
16. https://developer.x.com/en/docs/twitter-api/data-dictionary/object-model/user
17. https://www.mongodb.com/docs/manual/core/indexes/index-types/index-text/
18. https://www.mongodb.com/docs/manual/reference/operator/query/
19. https://developer.x.com/en/docs/twitter-api/data-dictionary/object-model/place
20. https://www.mongodb.com/docs/manual/geospatial-queries/
21. https://www.mongodb.com/docs/atlas/atlas-search/
22. https://www.mongodb.com/docs/atlas/atlas-vector-search/vector-search-overview/
23. https://www.mongodb.com/docs/manual/
24. https://www.mongodb.com/docs/manual/reference/operator/aggregation/
25. https://www.mongodb.com/docs/manual/core/aggregation-pipeline-optimization/
26. https://www.mongodb.com/docs/manual/core/views/create-view/
27. https://www.mongodb.com/docs/atlas/atlas-stream-processing/overview/
28. https://www.mongodb.com/docs/compass/current/query-with-natural-language/prompt-natural-language-query/
29. https://www.mongodb.com/docs/manual/reference/operator/query/regex/
30. https://www.mongodb.com/docs/upcoming/reference/explain-results/
31. https://www.mongodb.com/docs/generative-ai-faq
32. https://learn.microsoft.com/en-us/legal/cognitive-services/openai/data-privacy

## References


Bradshaw, Shannon, Eoin Brazil, and Kristina Chodorow. 2019. *MongoDB: The Definitive Guide: Powerful and Scalable Data Storage.* 3rd edition. Beijing, China; Sebastopol, CA: O'Reilly Media.

DeBarros, Anthony. 2018. *Practical SQL: A Beginner's Guide to Storytelling with Data.* San Francisco: No Starch Press.

MongoDB, Inc. n.d. "What Is MongoDB? - MongoDB Manual v7.0." Accessed July 3, 2024. https://www.mongodb.com/docs/manual/.

PostgreSQLTutorial. n.d. "PostgreSQL Tutorial." PostgreSQL Tutorial. Accessed July 3, 2024. https://www.postgresqltutorial.com/.

The PostgreSQL Global Development Group. n.d. "PostgreSQL: Documentation." Accessed July 3, 2024. https://www.postgresql.org/docs/.

# 6

# *Data Visualization*

## Data Visualization Basics

Data analytics aims to identify patterns and obtain insights for action planning or decision-making from collected data. While querying data can answer some questions, not all patterns can be identified with queries—one example is Anscombe's quartet (Anscombe 1973), which contains four groups of datasets, as shown in Table 6.1.

When we apply different aggregation functions to compute the means and variances on the four datasets, we obtain the same results. Similarly, if we calculate the correlation coefficients or establish a linear regression model, the correlations and model parameters are identical. Statistically, this implies that the four datasets exhibit the same relationships. However, plotting the datasets on a chart reveals different patterns. This visual representation highlights variations and nuances that are not apparent from statistical summaries alone, emphasizing the importance of visual data analysis in uncovering underlying differences. Figure 6.1 displays four scatter plots of those datasets: X1 and X2 have a positive linear relationship; X2 and Y2 have a nonlinear relationship; X3 and Y3 have a stronger linear relationship with one outlier; X4 and Y4 have no linear relationship. Anscombe's quartet proves that data visualization is not a visual decoration but an essential component in data analytics.

In addition, data visualization decodes data with *preattentive attributes*,[1] such as length, width, shape, size, position, and color. Our brains process such information at an extremely high speed, faster than processing words or numbers. Visualization also facilitates communication due to the *picture superiority effect*: humans retain information from pictures or images longer than words (Hockley 2008). Therefore, data visualization is an effective and efficient approach to data analysis that can accelerate the analysis process and help clients appreciate results.

Business intelligence (BI) tools like Amazon QuickSight,[2] Tableau,[3] or Power BI[4] can visualize data and create interactive dashboards. ArcGIS Pro[5] is a geographic information system (GIS) application designed to analyze and visualize spatial data and create professional maps.

DOI: 10.1201/9781003437611-6

**TABLE 6.1**
Statistics of Anscombe's Quartet (Data source: Anscombe 1973)

| | Dataset1 | | Dataset2 | | Dataset3 | | Dataset4 | |
|---|---|---|---|---|---|---|---|---|
| | X1 | Y1 | X2 | Y2 | X3 | Y3 | X4 | Y4 |
| | 10 | 8.04 | 10 | 9.14 | 10 | 7.46 | 8 | 6.58 |
| | 8 | 6.95 | 8 | 8.14 | 8 | 6.77 | 8 | 5.76 |
| | 13 | 7.58 | 13 | 8.74 | 13 | 12.74 | 8 | 7.71 |
| | 9 | 8.81 | 9 | 8.77 | 9 | 7.11 | 8 | 8.84 |
| | 11 | 8.33 | 11 | 9.26 | 11 | 7.81 | 8 | 8.47 |
| | 14 | 9.96 | 14 | 8.1 | 14 | 8.84 | 8 | 7.04 |
| | 6 | 7.24 | 6 | 6.13 | 6 | 6.08 | 8 | 5.25 |
| | 4 | 4.26 | 4 | 3.1 | 4 | 5.39 | 19 | 12.5 |
| | 12 | 10.84 | 12 | 9.13 | 12 | 8.15 | 8 | 5.56 |
| | 7 | 4.82 | 7 | 7.26 | 7 | 6.42 | 8 | 7.91 |
| | 5 | 5.68 | 5 | 4.74 | 5 | 5.73 | 8 | 6.89 |
| *Mean* | 9 | 7.5 | 9 | 7.5 | 9 | 7.5 | 9 | 7.5 |
| *Variance* | 11 | 4.127 | 11 | 4.127 | 11 | 4.127 | 11 | 4.127 |
| *Correlation* | 0.816 | | 0.816 | | 0.816 | | 0.816 | |
| *Regression* | Y = 3 + 0.5*X | | Y = 3 + 0.5*X | | Y = 3 + 0.5*X | | Y = 3 + 0.5*X | |

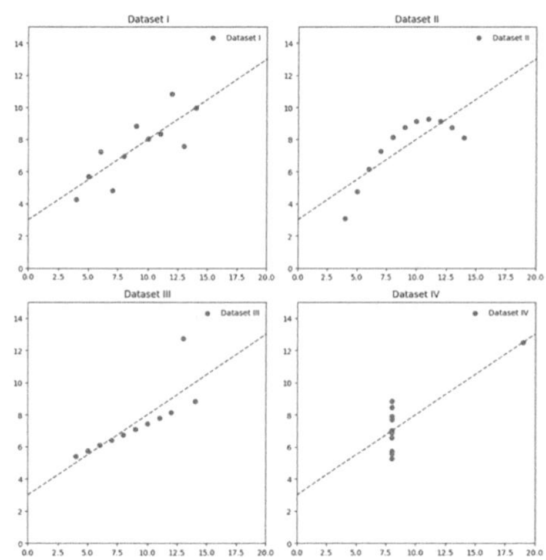


**FIGURE 6.1**
Visualization of Anscombe's Quartet (Data source: Anscombe 1973, Plotted by ChatGPT)

This chapter introduces data visualization basics and presents AI-powered data visualization case studies. To learn more about visualization theories, we recommend *Data Visualization: A Successful Design Process* (Kirk 2012). *Cartography: Thematic Map Design* (Dent, Torguson, and Hodler 2008) is an excellent book about visualizing spatial data. *Beautiful Visualization* (Steele and Iliinsky 2010) and Tableau Viz Gallery[6] provide great examples to inspire your visualization projects.

## Types of Charts

The most common charts include *histograms, box plots, scatter plots, line charts, bar charts,* and *pie charts.* A histogram displays a *statistical distribution* of a single variable. The histogram in Figure 6.2 shows the distribution of the states' median household incomes in 2019. The X-axis shows *bins,* e.g., the different income ranges; and the Y-axis displays the number of records in each bin, such as the number of states in each income range. Most states' median household income is around $60,000 and only one below 45,000.

We can use a box plot to compare the statistical distributions of multiple variables. A box plot shows a *percentile summary* of a dataset. The lower and upper whiskers represent the minimal and maximum values, excluding outliers. The box starts from the lower to the upper quartile, covering 50% of the dataset. The middle line of the box represents the median value. The dots beyond the whiskers are outliers, with values less than 1.5 times the lower quartile or more than 1.5 times the upper quintile. The box plot in Figure 6.2 shows the distribution of state incomes over multiple years. All states' income gradually increased from 2005 to 2019, except for one outlier, which was always around $20,000 . However, the income difference among the states became larger after 2016.

Scatter plots are designed to explore relationships between two continuous variables. The common practice is to place dependent variables on the Y-axis and independent variables on the X-axis. This arrangement helps in visualizing how changes in the independent variable influence the dependent

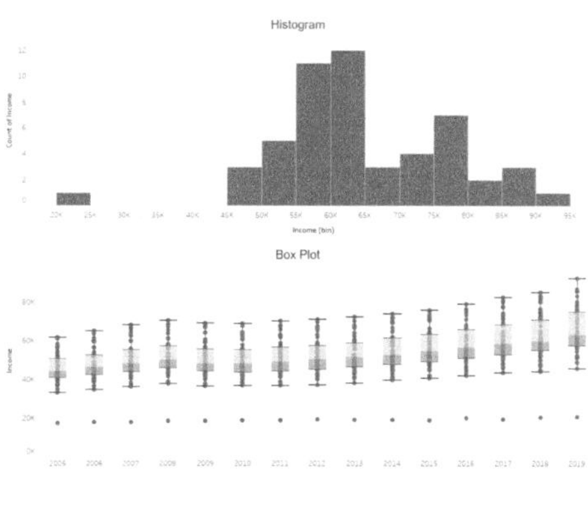


**FIGURE 6.2**
Histogram and Box Plot

variable, making it easier to identify trends, correlations, and potential outliers in the data. The axis scaling can be *arithmetic, semi-logarithmic (semi-log),* or *log-log* scaling.

- Arithmetic scaling: The spacing of the tick marks on both axes is constant, and equal spacing represents equal values. The origins of the axes start at zero or the beginning of the data. We can see a straight line when the independent variable increases at a constant *amount* with the dependent variable.
- Semi-log scaling: Arithmetic scaling is applied on the X-axis, and logarithmic scaling on the Y-axis. We can see a straight line when the independent variable increases at a constant *rate* with the dependent variable. A steeper slope of the line indicates a more significant change rate.
- Log-log scaling: The logarithmic scaling is applied on both axes to show the rate change of one variable against another.

The scatter plot in Figure 6.3 displays the relationship between states' populations and median household incomes in 2019. The chart uses arithmetic scaling and indicates no relationship between the population and the income at the state level.

A line chart typically depicts a *trend* of variables over time. We place temporal data, such as date and time, on the X-axis. For example, the line chart in Figure 6.3 shows the change in median household incomes in Maryland and Virginia. Maryland's income was constantly higher than Virginia's over the past years, and both states' incomes gradually increased except for a sudden drop in 2009.

Both bar charts and pie charts can compare quantities in different categories. A bar chart utilizes heights or lengths, one-dimensional symbols, to represent values. A pie chart represents values with sizes of pies that are two-dimensional symbols. However, people tend to *underestimate* the values

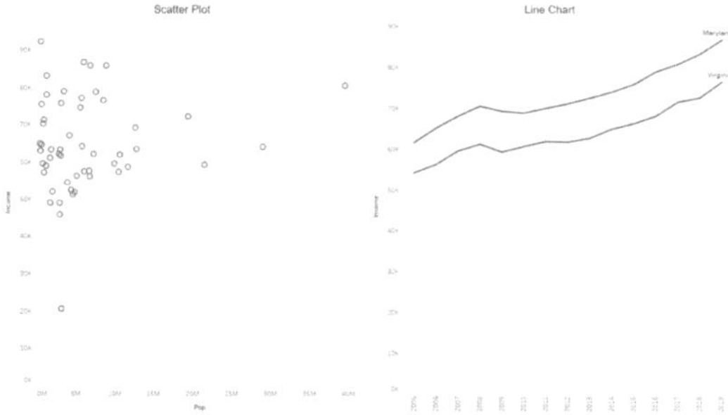


**FIGURE 6.3**
Scatter Plot and Line Chart

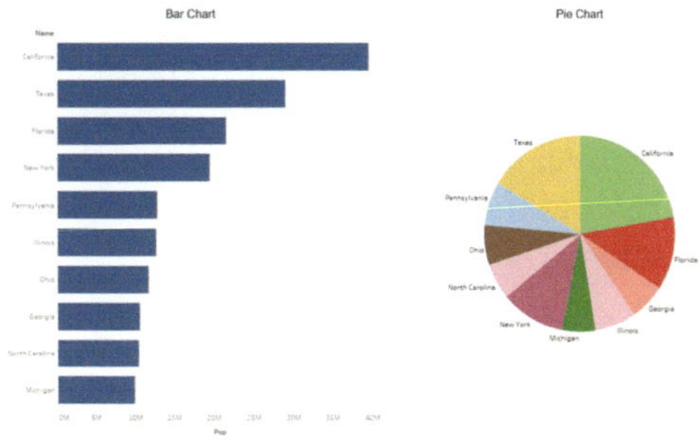


**FIGURE 6.4**
Bar Chart and Pie Chart

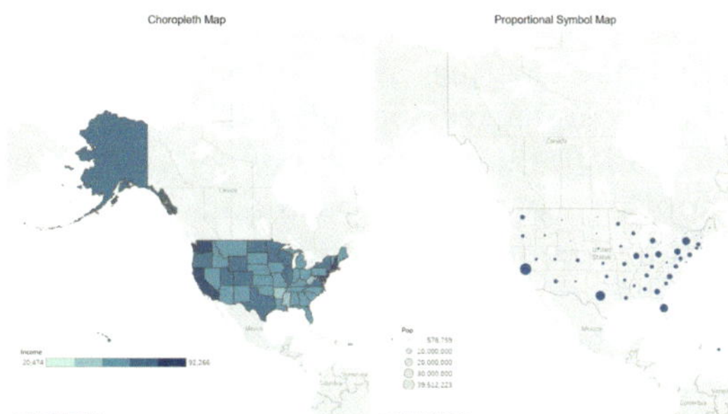


**FIGURE 6.5**
Choropleth Map and Proportional Symbol Map

represented by two-dimensional symbols.[7] Therefore, pie charts are *not* recommended. Figure 6.4 shows how a bar chart and a pie chart display the 2019 population of the top 10 states. It is clear to identify that Florida has more pollution than New York on the bar chart, but not easily on the pie chart. Besides, the pie chart uses multiple colors to distinguish different states. Without labels, people may be unable to tell the difference between North Carolina and Illinois. For bar charts, we should avoid using too many bars to create a *Moiré Effect*[8] that can cause discomfort to many viewers.

Many BI tools can create simple maps, such as *choropleth* maps or *proportional symbol* maps. The choropleth map, or a filled map, uses *colors* to represent the values across geographic spaces. As a best practice, choropleth maps visualize *ratios* or *normalized* data, such as population densities or median household incomes. The choropleth map in Figure 6.5 shows the 2019 median household incomes in different states. Higher-income states are generally distributed on the west or east coast.

The proportional symbol map depicts the values with different-*sized* symbols. The proportional symbol map in Figure 6.5 shows the 2019 population distribution in the USA. As we mentioned before, viewers may underestimate the value of those circles. We can utilize the *Appearance compensation*[9] in professional GIS software to counteract such perception.

BI tools or GIS software can create variants of the above charts or other types of maps. Please refer to the recommended textbooks to explore the different visualization options.

## Colors

Colors are essential in data visualization, but people's perceptions of colors are subjective. For example, yellow may indicate cheerfulness but also represent hate or warning. We may feel passion or excitement when seeing a red color, but we may also feel anger or danger. Therefore, colors can *neutrally* highlight or emphasize specific data points and convey *particular* messages. In addition, the perception of colors is relative rather than absolute. Color appearance depends on the background or surrounding colors (Brown and MacLeod 1997). That is *why you should never trust a data visualization.*[10]

There are *additive* color theory and *subtractive* color theory. The additive color theory describes how *lights* produce color. Red, green, and blue are the primary colors. Combining two additive primary colors produces magenta, cyan, or yellow, and combining all of them produces white. Colors produced by *printing* are based on the subtractive color theory, and the primary colors include magenta, cyan, and yellow. Combining two subtractive primary colors produces red, green, or blue, and combining all produces black.

The components of color include *hue, saturation,* and *value.* Hue refers to the name of various colors, such as red, blue, etc. Saturation indicates the intensity or purity of a color compared to a neutral gray. Value describes the lightness or darkness of a color. In addition, we can choose *sequential, diverging,* or *qualitative* schemes in data visualization. Sequential schemes use a single hue with variant saturations or values to depict the change of quantitative values. For example, the choropleth map in Figure 6.5 uses a sequential scheme to compare income values. The diverging schemes join sequential schemes with two hues to highlight the minimal and maximal values. The qualitative schemes use multiple hues to distinguish various categories. The pie chart in Figure 6.4 uses a qualitative scheme to represent different states.

In addition, we should avoid using a *red-green* color combination in data visualization. Approximately 1 in 12 men and 1 in 200 women[11] have *color vision deficiency* (CVD) that can't distinguish red and green. We can use a colorblind-friendly palette[12] when possible; a *blue-orange* color combination is also a good alternative.

## Design Principles

Visualization design involves science and art; good designs follow the same principles. In the chapter "Graphical Excellence," Tufte (2001) concludes that graphical excellence is a "well-designed presentation of interesting data – a matter of substance, of statistics, and of design" that "gives the viewer the greatest number of ideas in the shortest time with the least ink in the smallest space." Shaffer[13] also summarized the *4Cs* principle of data visualization, i.e., *clear, clean, concise,* and *captivating.*[14]

Specifically, we should choose the right type of chart or map to answer our questions effectively. The previous section introduces the basic types of charts and maps. We recommend the Tableau white paper, *Which Type of Chart or Graph is Right for You* (Tracy Rodgers, n.d.),[15] to help select appropriate charts. In addition, we should place the most crucial chart at the top or top left of the dashboard to accommodate viewers' scanning patterns.[16] If we use symbols of different colors or sizes, we must provide *legends,* which should be close to the associated charts or maps.

The *KISS* principle (keep it simple, stupid) also applies to data visualization. Ideally, the audience can understand the basic ideas of a dashboard in *five seconds.*[17] Therefore, we should limit the charts or colors used in a single dashboard to reduce viewers' cognitive load. Generally, no more than five charts or colors should be used. When different colors are required, we should use them consistently.

*Fonts* also play an essential role in visualization design, as different fonts are associated with certain personalities (Shaikh, Chaparro, and Fox 2006). As Pidsley (2022) suggested, we should use fonts from the same font families but avoid using fonts that are too similar or too different. We should also use no more than *three* different fonts and ensure that all the texts are *legible.* For more tips on using fonts, please read Chris Pidsley's blog.[18]

Finally, providing *interactivity such as* filters, tooltips, or actions becomes vital in modern dashboards. Users are most likely viewing dashboards with digital devices and expect to interact with them using their fingers. Interactivity allows users to explore the data further and obtain information not directly displayed on the dashboard. Please read Tableau Visual Best Practices (Salesforce, Inc., n.d.)[19] for learning how to create interactive visualizations.

## Data Visualization with AI

This section demonstrates case studies of data visualization using Python and business intelligence tools with the assistance of Generative AI. The

BI tools include Amazon QuickSight[20] and Atlas Charts[21]; and the GenAI assistants are Jupyter AI[22], Amazon Q[23], and MongoDB Natural Language Charts.[24] When we wrote this book, many AI tools were in preview or just released, and none provided academic licenses. Therefore, we use the paid or preview versions of those AI tools in this section.

## Visualize Data in Python with Jupyter AI

Once data is collected and stored in a database, we can follow the examples in Chapter 3 to analyze and visualize the data in Python. Figure 6.6 illustrates a typical analytical process:

- We can use the database APIs, such as *psycopg2*[25] or *pymongo*,[26] to load the data from the database into Python. We also use query languages, like *SQL* or *MongoDB CRUD*, to filter or aggregate the datasets before loading them into Python.
- When data is loaded in Python, we can analyze and visualize it using data analytical libraries, such as *pandas*[27] and *matplotlib*.[28] Although pandas can filter and aggregate data, we recommend processing data in databases to reduce the computational load on JupyterLab instances and *prevent unnecessary data movement*.[29]
- With Jupyter AI (Project Jupyter, n.d.), we can bring Generative AI capabilities to assist data analysis and visualization in Python. In previous chapters, we used Amazon Q Developer[30] in Cloud9 to

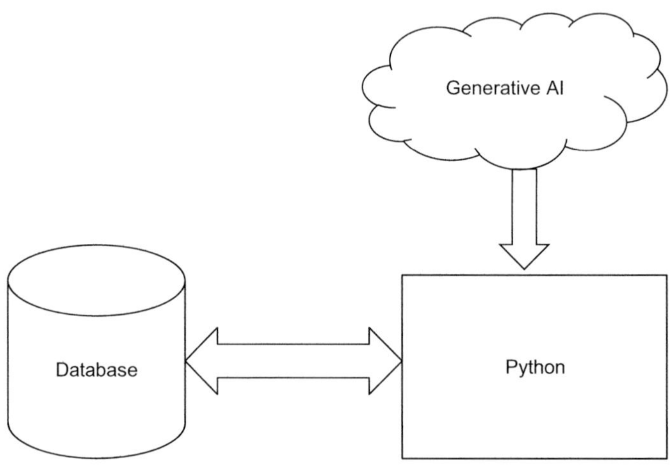


**FIGURE 6.6**
Data Visualization in Python with AI

write Python codes and chatted with ChatGPT[31] for SQL suggestions. Jupyter AI can work with different Python editors, in particular JupyterLab and Google Colab,[32] and support various Generative AI model providers, including OpenAI, Hugging Face,[33] Gemini,[34] etc. Therefore, we use Jupyter AI in this section because it supports a wide range of products.

This section uses the collected Census data and chooses OpenAI GPT-4[35] as the GenAI model. Before we start, make sure the PostgreSQL and JupyterLab instances are started. We must also purchase an OpenAI API from OpenAI[36] and store the API key in AWS Secrets Manager, as we did in Chapter 3.

In JupyterLab, we can create a new notebook named *Analyze_Census_Data.ipynb*. In the notebook, we need to install the following Python packages:

- jupyter-ai[37]: the JupyterLab extension to interact with Generative AI models
- langchain-openai[38]: the LangChain package to call OpenAI APIs

```
pip install jupyter-ai
pip install langchain-openai
```

We may need to restart the kernel after installing the packages by clicking the *Restart the Kernel* button on top of the notebook. Next, we reuse the *Secret Manager function* and the *Connect to Database* codes from the *Collect_Census_Data*.ipynb. You can also find those codes in the textbook's GitHub repository.[39] We then pass the *OpenAI API key* to the *environment variables* that store configuration settings:

```
import os
os.environ["OPENAI_API_KEY"] = get_secret('openai')
['api_key']
```

Next, we load the *Jupyter AI magic commands*.[40] We use those commands to manage various GenAI models and pass our prompts to the models for coding assistance. Please check the *Jupyter AI GitHub* repository for more example notebooks.[41]

```
%load_ext jupyter_ai_magics
```

For example, we can use the *%ai* command to list the available models:

```
%ai list
```

The output lists all the supported GenAI model providers. It shows that the *openai* models are ready to use, as we just installed the required Python libraries and provided the API key. Please check the JupyterAI website for a complete list of the model providers.[42]

We are now ready to load the Census data and visualize it in Python. Suppose we want to see the change in the income data in Virginia. We first need to define the SQL query to filter the exact data. We could JOIN the *name* table with the *income* table and use the WHERE clause to filter only the data in Virginia. Fortunately, we already have the *name_income* VIEW that joins the *name* table and the *income* table. So, alternatively, we SELECT the data from the *name_income* view and use the WHERE clause to filter the data in Virginia. We choose to use *fips* in the WHERE clause because it is a *primary key* that is indexed automatically.

```
sql = """
    select * from name_income
    where fips = '51'
"""
```

Next, we can use *pandas* to load the data with the defined SQL query and pass it to a *DataFrame* object.

```
import pandas as pd
df = pd.read_sql(sql, con=conn)
df.head() #show the top 5 rows
```

Code output:

| | fips | name | income | year |
|---|---|---|---|---|
| 0 | 51 | Virginia | 54240 | 2005 |
| 1 | 51 | Virginia | 56277 | 2006 |
| 2 | 51 | Virginia | 59562 | 2007 |
| 3 | 51 | Virginia | 61233 | 2008 |
| 4 | 51 | Virginia | 59330 | 2009 |

We can now create a line chart to show the change in income in Virginia over the past years:

```
df.plot(x='year', y='income')
```

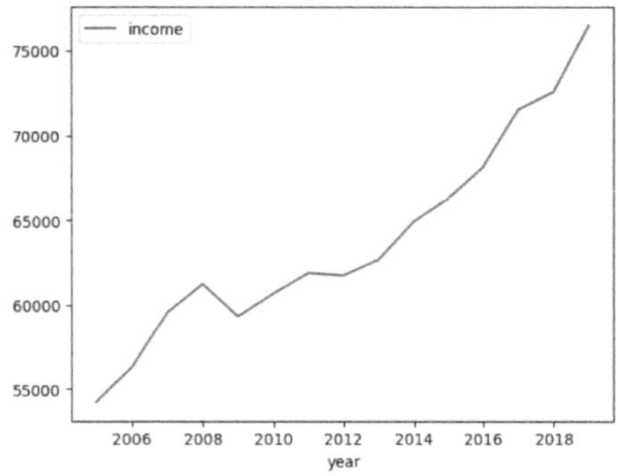


**FIGURE 6.7**
VA Income over Years

A line chart is now created. As shown in Figure 6.7, we can identify a pattern in which the median household income in Virginia gradually increased since 2005, with exceptions in 2009 and 2012.

To use GenAI assistants, we can call the *%%ai* magic command followed by a specific GenAI model name. Beneath the model's name, we can define our prompt in natural language to ask questions. For example, we can ask the *GPT4* model to explain a specific Python code by providing the notebook cell number in brackets:

```
%%ai gpt4
  explain the code in {In[8]}
```

The OpenAI GPT4 model analyzes the code in Cell 8, i.e., the code loading data into DataFrame, and provides the following explanation. Most of the content is accurate, except that the *conn* variable is a *PostgreSQL* connection object. Without reading the previous codes, the model is not supposed to know the exact engine type.

```
The code is written in Python and uses the pandas
library for data manipulation and analysis.
```

import pandas as pd – This line is importing the pandas library and giving it the alias "pd". This allows you to use the shorthand "pd"" instead of writing "pandas" when calling pandas methods.

df = pd.read_*sql(sql, con=conn) – This line uses the pandas method read*_sql to read an SQL query or database table into a DataFrame. The sql variable is the SQL query you want to execute, and con is the SQLite connection object that represents the database you want to execute the query on. The result of this query is assigned to the variable df.

```
df.head() - This line displays the first 5 rows of
the DataFrame df. The head() function is used to get
the first n rows, with 5 as the default number. If you
wanted to display the first 10 rows for example, you
would use df.head(10).
```

We can pass Python variables to the GenAI model to write codes using the variables. For example, in the following prompt, we ask AI to use the data loaded in the *df* variable to calculate the year-over-year growth rate and create a bar chart to visualize the result. We specify the output format in Python code and enclose the variable name in curly brackets:

```
%%ai gpt4 -f code
Calculate the year-over-year income growth rate in
{df}
and visualize it on a bar chart
```

Below is the AI-generated code. It loads the *df* data into a *data* variable and calculates the growth rate with a *pct_change* function.[43] It then creates a bar chart with the results. The model also understands that the data is from Virginia and provides an appropriate chart title. The GenAI model also adds comments to explain the Python code.

```
import pandas as pd
import matplotlib.pyplot as plt
# Create a dataframe
data = {
    'fips': [51]*15,
    'name': ['Virginia']*15,
    'income': [54240, 56277, 59562, 61233, 59330,
    60674, 61882, 61741, 62666, 64902, 66262, 68114,
    71535, 72577, 76456],
'year': [2005, 2006, 2007, 2008, 2009, 2010, 2011,
2012, 2013, 2014, 2015, 2016, 2017, 2018, 2019]
}
df = pd.DataFrame(data)

# Calculate year-over-year income growth rate
df['YoY_growth'] = df['income'].pct_change() * 100
```

```
# Create a bar chart
plt.figure(figsize=(10, 6))
plt.bar(df['year'][1:], df['YoY_growth'][1:]) #
Exclude the first year as it has no previous year to
compare to
plt.xlabel('Year')
plt.ylabel('YoY Income Growth Rate (%)')
plt.title('Year-over-Year Income Growth Rate in
Virginia')
plt.show()
```

After executing the generated codes, a bar chart is shown in Figure 6.8. The Virginia income growth rates were negative in 2009 and 2012, consistent with the results we calculated in Figure 6.7.

We now try to resolve another question with the GenAI model, from loading the data to visualizing it. We first ask the model to load the data from the database into Python and instruct it to use the existing *conn* object and the *name_income* view:

```
%%ai gpt4 -f code
We have a view named name_income in an SQL database
you don't need to redefine the connection string
use the {conn} defined in {In[3]} to load all the
recored into pandas dataframe
```

The generated code still creates a new *connection* and a *cursor* object and authorizes those objects with our *get_secret* function. The code then uses the cursor to fetch all the rows with a correct SQL statement and convert the data into a *pandas* DataFrame. It also passes the column names to the DataFrame and closes the connection and cursor.

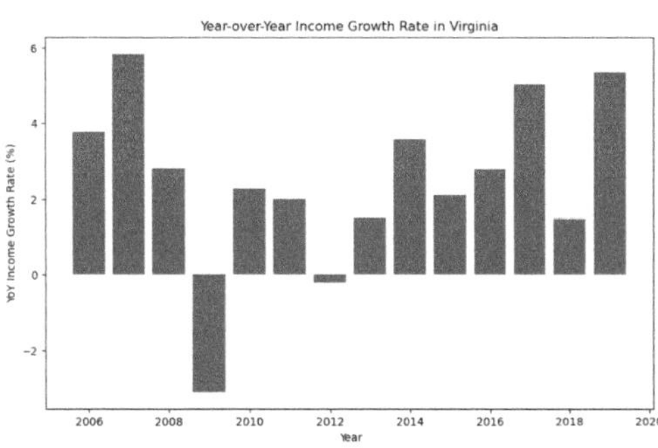


**FIGURE 6.8**
Year-over-Year Income Growth Rate in Virginia

```
import pandas as pd
import psycopg2

# Define connection string
conn = psycopg2.connect(host=get_secret('postgresql'
)['host'],
    user=get_secret('postgresql')['username'],

password=get_secret('postgresql')['password'],
    dbname=get_secret('postgresql')['engine'])

# Create a cursor object
cur = conn.cursor()

# Execute the SQL query
cur.execute("SELECT * FROM name_income")

# Fetch all results from the cursor into a list
rows = cur.fetchall()

# Get the column names from the cursor description
columns = [desc[0] for desc in cur.description]

# Create a pandas DataFrame from the results
df = pd.DataFrame(rows, columns=columns)

# Close the cursor and connection
cur.close()
conn.close()
```

We then ask AI to tell us what information is in the imported data:

```
%%ai gpt4
What is the information in {df}
```

The model provides the following responses that are also accurate:

```
This data represents the average income for each US
state and Puerto Rico for the years 2005 to 2019.

The columns represent:
```

```
fips: A code which is a combination of two-digit
country code and two-digit state code.

name: The name of the state

income: The average income for that state in the
given year

year: The year in which the income data was
collected.
```

Finally, we ask AI to calculate each state's average year-over-year income growth rate and visualize the result.

```
%%ai gpt4 -f code
Calculate the average year-over-year income growth
rate for each state in {df}
and create a chart to show the result
```

Below is the generated code. It sorts the DataFrame by state names and years, calculates the growth rate for each state, and then computes the average growth rate per state. The code finally creates a bar chart to show the results and defines the chart style, like selecting a horizontal bar and specifying the chart size and color.

```
import pandas as pd
import matplotlib.pyplot as plt

# Assuming the data is in a DataFrame named df
df['income'] = df['income'].astype(float)
df['year'] = df['year'].astype(int)

# Group by state and calculate the average annual
growth rate
df = df.sort_values(['name', 'year'])
df['income_growth'] = df.groupby('name')['income'].p
ct_change() + 1
df = df.groupby('name')['income_growth'].agg('mean')
- 1

# Create a bar chart
df.sort_values().plot(kind='barh', figsize=(10, 20),
color='skyblue')
```

```
plt.title('Average Year-Over-Year Income Growth Rate
by State')
plt.xlabel('Average Year-Over-Year Growth Rate')
plt.ylabel('State')

plt.show()
```

The created chart is in Figure 6.9, with D.C. having the highest average growth rate and Puerto Rico having the lowest rate.

While analyzing data with AI is like magic, we should always double-check and modify the AI-generated codes when necessary. For example, we asked the model to generate the income growth bar chart multiple times, and each time, the model created slightly different versions. One of them mistakenly reported the average income instead of the growth rate. Therefore, a basic understanding of Python and data analytics is still required to use those GenAI models. The performance of different models also varies. For example, OpenAI *gpt4*[44] creates better content than *gpt3*.[45]

In addition, if we want GenAI models to provide more relevant coding assistance, we have to allow models to access our data. If datasets contain sensitive information, we must choose a model that runs locally or in our private cloud to comply with data security and privacy regulations.

After finishing the analysis, please push the new codes to the GitHub repository for future reference and stop the database and notebook instances to save the AWS credits.

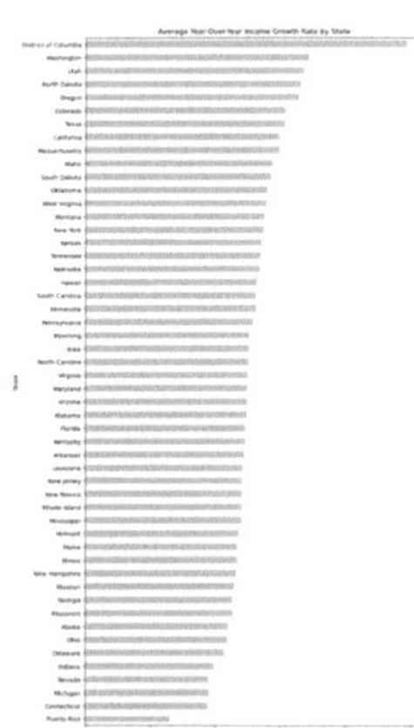

**FIGURE 6.9**
Average Year-over-Year Income Growth Rate by State

## Visualize Data in QuickSight with Amazon Q

Once the BI tools connect to data databases, users can drag and drop data fields to create various charts. The BI tools convert users' actions into SQL commands, send the queries to databases, and display the received data in charts. BI tools have friendly user interfaces, easy-to-use functions, and support interactive online visualizations.

As we wrote this textbook, many BI tools have just introduced Generative AI capabilities into their BI system. For example, on April 30, 2024, Amazon Q became generally available in Amazon QuickSight.[46] Unfortunately, none of the AI-powered BI tools provides academic licenses. Therefore, this section uses a paid QuickSight version. As shown in Figure 6.10, we use BI tools like QuickSight to connect to databases, such as the PostgreSQL instance. We can create charts in QuickSight manually or ask Amazon Q,[47] the Amazon Generative AI assistant, to modify existing charts, perform new calculations, or design new visualizations.

Once we log in to QuickSight, we can create *datasets, analyses, dashboards, topics,* and *stories.* Topics and stories are new features that utilize GenAI to generate narrative summaries or drafts from datasets. First, we must build and connect a dataset to our PostgreSQL database. QuickSight supports various data sources, such as file uploading, data lakes, data warehouses, or databases. Figure 6.11 shows how to connect to a PostgreSQL server: We need to provide the host URL, database name, username, user password, and port number, similar to how we use pgAdmin to connect our database instance.

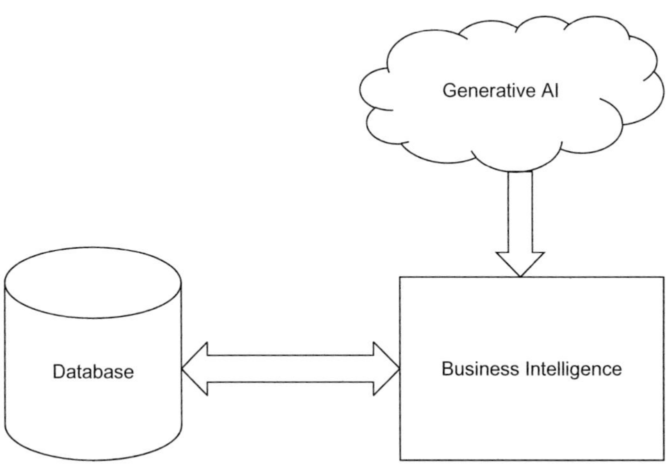


**FIGURE 6.10**
Visualize Data with AI-powered BI Tools

**FIGURE 6.11**
Connect QuickSight to PostgreSQL

**FIGURE 6.12**
Join Tables in QuickSight

After establishing the connection, we can visually edit table relationships. For example, we can join the *name* and *population* tables based on common *fips* fields and define the appropriate *join type*, as shown in Figure 6.12. QuickSight supports the four types of joins introduced in the previous chapter and can join tables based on multiple fields. We can also use *views* in databases if they have been created.

Data clearing in BI tools is simple as well. For example, we exclude the duplicated *fips* field, rename the *name* field to the *state*, and specify the data type as *geospatial*. QuickSight can create maps based on latitude and longitude and recognize common places, such as country, state, county, city, and zip code. We also create a *calculated field* to store the *year* value in date format and rename the pop field to *population*. Figure 6.13 shows the cleaned data in QuickSight. The graphic interface makes it much easier to manage datasets, and we don't need to write complicated SQL statements to join or clean them.

Two query modes are available on QuickSight: *direct query* or *SPICE* (Super-fast, Parallel, In-memory Calculation Engine).[48] In the direct query, all the queries are executed in databases, and QuickSight communicates with databases to retrieve the results. If we select the SPICE mode, data will be loaded to the SPICE in-memory engine for rapid data query and process. Please note: a 10 GB SPICE is allocated to each QuickSight author license; additional SPICE storage causes extra charges.[49]

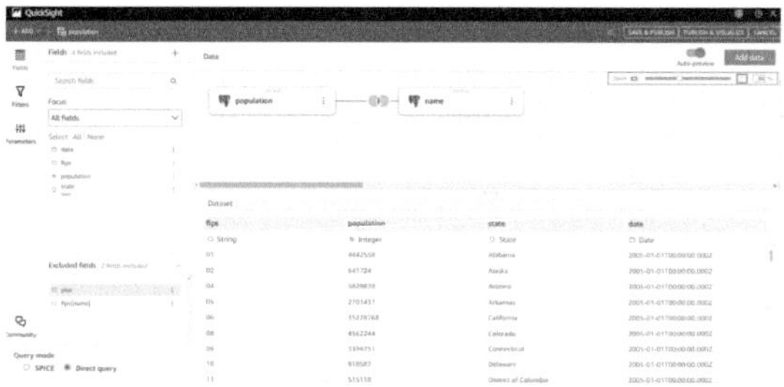

**FIGURE 6.13**
Cleaned Data in QuickSight

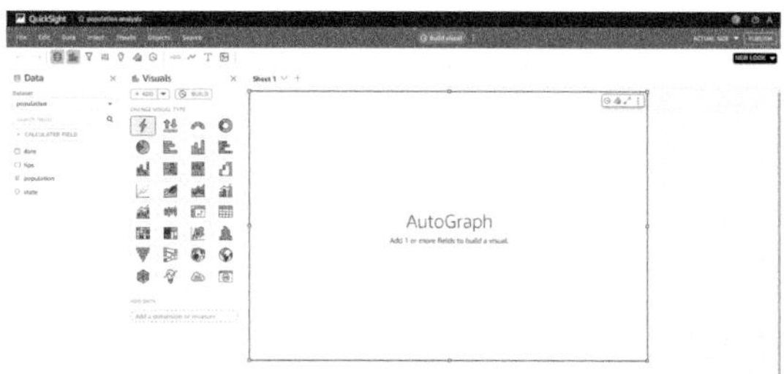


**FIGURE 6.14**
QuickSight Analysis Interface

We can keep the direct query and start a new *Analysis* with the created dataset. Figure 6.14 shows the interface of QuickSight Analysis: We can choose the fields on the *Data* panel, select a chart type on the *Visuals* panel, and define *aggregation* functions or change chart *properties*. A chart shows up on the right. We can provide *interactivity* or click the *Build with Q* button to let AI edit existing charts or build new charts.

For example, we first select the *state* and *population* fields. A bar chart is automatically generated to show the *sum* of the population by state. We can change the chart type to a *proportional symbology map* and modify the aggregation function to show each state's *maximum* population. Figure 6.15 displays the created map, where California has the most giant circle due to its population. On top of the map, we can click the *Build with Q* button to modify the charts with prompts.

Similarly, we can add a second chart and select the year and population field on the Data panel. A *line chart* is automatically created to show the total population over time. We can change the granularity of the date to *year* on the Visual panel and click the three dots on the chart to *add a forecast*. Figure 6.16 shows the line chart with a yellow forecast of the USA population. The chart indicates that the US population reached 300 million in 2006 and is forecasted to reach 343 million in 2024, with an upper bound of 345 million and a lower bound of 340 million. According to the US Census Bureau, the

**FIGURE 6.15**
Max of Population by State

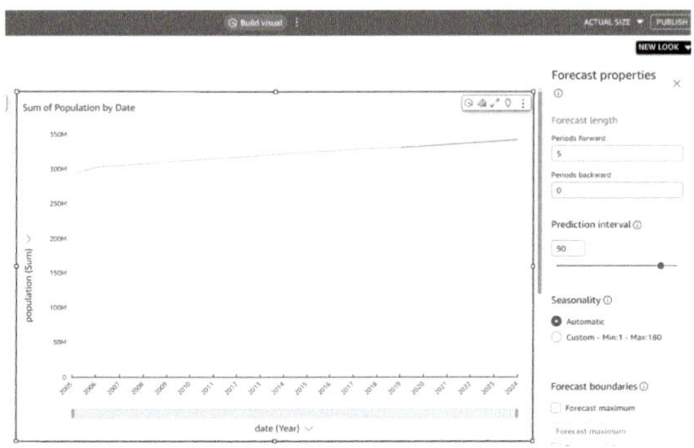


**FIGURE 6.16**
Sum of Population by Year

projected USA population on January 1, 2204, is 335,893,238,[50] lower than the forecasted lower bound value.

We can ask Amazon Q to edit an existing chart or create a new chart. For example, we asked Amazon Q to:

```
use a line chart to show the year-over-year
population difference in each state
```

We received the response in Figure 6.17. Amazon Q calculated the year-over-year population difference for each state and created a line chart to show the results. Each state is assigned a different color. We can *Add the chart to Analysis* with a single click or modify the prompt to create a new chart.

In addition, we can add *insights,* i.e., automatically generated narratives, to describe the data. For example, the automatically generated top and bottom movers are below. Users can choose other generated insights or create customized insights.

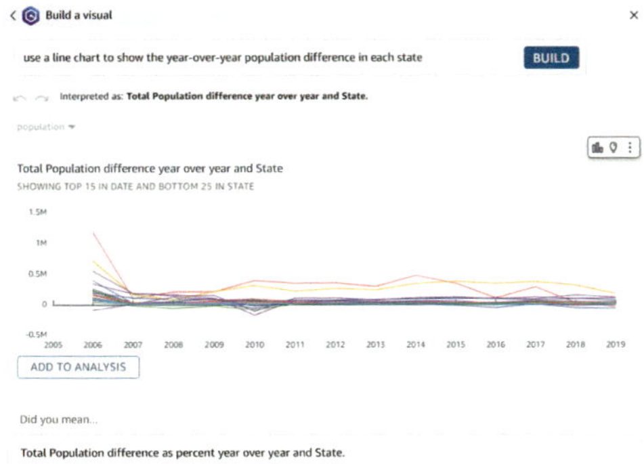


**FIGURE 6.17**
AI-Generated Charts in QuickSight

```
Top movers
Top custom aggregation of year over year Population
movers for 2019 are:
Kansas increased by 211.84% (3,417) from -1,613 to
1,804.
Rhode Island increased by 188.04% (4,370) from
-2,324 to 2,046.
Wyoming increased by 164.77% (2,600) from -1,578 to
1,022.

Bottom movers
Bottom custom aggregation of year over year
Population movers for 2019 are:
Mississippi decreased by 527.2% (-12,811) from 2,430
to -10,381.
Pennsylvania decreased by 432.96% (-6,594) from
1,523 to -5,071.
California decreased by 319.8% (-65,214) from 20,392
to -44,822.
```

QuickSight provides three interactive functions: *tooltips, filters,* and *actions.*

- Tooltips provide information about specific elements when the user hovers the cursor. Analysts can define the information format and content that appears in the tooltips.
- Filters can be applied to regular or calculated fields or on multiple charts to show relevant information. They can filter values like dates, texts, and numbers.

**FIGURE 6.18**
Interactive Dashboard in QuickSight

- Actions allow customers to explore the charts interactively. The actions can be filter actions that filter data by clicking elements on charts, navigation actions that enable switching sheets in an analysis, or URL actions that open web pages.

For example, we open the map *properties* panel, add a *filter action* to filter the state field, and apply the filter to *all charts*. We then publish the analysis as a dashboard. Figure 6.18 shows the created dashboard. When the user clicks California on the map, the selected state is highlighted, and all other charts show California information only.

We can share the dashboard with our colleagues or clients, who can explore interactive visualizations online. We can also build *data stories* or create *topics* to generate executive summaries of key insights. The dashboard will automatically update and notify the users when new data enters the database.

BI tools are an easy way to explore data visually and can share results interactively. However, most BI tools are closed resources and require expensive subscriptions. In addition, BI tools generally work best with structured data, such as relational databases or data warehouses. For example, if you want to use Tableau or QuickSight to visualize MongoDB data, you must do extra work, such as building a third query layer. In addition, the GenAI assistants, such as Amazon Q, can only perform simple tasks. For example, we tried to ask Amazon Q to change colors or filter states on the proportional symbol map, and it constantly reported errors or generated inaccurate visualizations.

## Visualize Data in Atlas Chart with OpenAI

In the previous sections, we used Python and BI tools to visualize data with Generative AI. Those Generative AI assistants, such as Jupyter AI

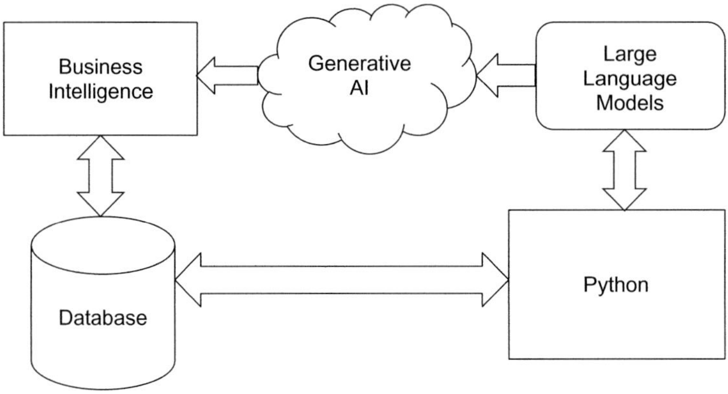


**FIGURE 6.19**
Data Analytics with LLM

and Amazon Q, are powered by large language models (LLM) as OpenAI and Gemini.[51] Alternatively, we can use LLM directly to analyze data. As shown in Figure 6.19, we can use Python to retrieve data from a database and call the LLM to examine the retrieved data. We then store the results in the database and visualize them using an AI-powered BI tool. This section uses OpenAI to analyze the collected Tweets and store the results in the MongoDB database. We can visualize the results using Atlas Charts,[52] a BI tool optimized for the MongoDB database. The Natural Language Charts[53] is an AI-powered visualization function released on March 11, 2024[54]. It is still a preview feature when we wrote this book.

We can start the JupyterLab instance and create a new notebook named *Analyze_Twitter_Data.ipynb*. We need to install the following Python libraries:

- pymongo[55]: manage the MongoDB database
- openai[56]: official OpenAI library to call OpenAI APIs

```
pip install pymongo
pip install openai
```

We can reuse the *Secret Manager Function* codes in the *Collect_Twitter_Data .ipynb*. You can find those codes in the textbook's GitHub repository.[57] After that, we need to import the Python libraries and load the MongoDB credentials and OpenAI API key.

```
import pymongo
from pymongo import MongoClient
import json
from pprint import pprint
```

```
from tqdm.auto import tqdm
import re

openai_api_key = get_secret('openai')['api_key']

mongodb_connect = get_secret('mongodb')['connect
ion_string']
```

To connect the MongoDB *demo* database, we can reuse the *Connect to the MongoDB cluster* code in the *Collect_Twitter_Data*.ipynb. Next, we extract the Tweets we are interested in and use Compass *Natural Language Query* to write the queries. For example, in Compass, we can write the following prompt:

```
Filter tweets with more than 100 retweets; only
return tweet text and tweet ID
```

Generative AI creates a document query, which we can export to our Python code using the *Export Query to Language* tool. Below are the codes exported from Compass. We can use them in our notebooks. You can always filter the Tweets you are interested in by using different prompts or queries. Please note: MongoDB Compass creates a client variable with an explicit connection string in the original exported query. Because we use AWS Secrets Manager to pass the connection string, we need to delete the connection string in the first line of the exported query. Also, we named the MongoDB client object *mongo_client*, so we need to rename the client name in the exported query.

```
filter={
    'tweet.public_metrics.retweet_count': {
        '$gt': 100
    }
}
project={
    'tweet.text': 1,
    'tweet.id': 1
}
#rename the client to mongo_client
result = mongo_client['demo']['tweet_collection']
.find(
    filter=filter,
    projection=project
)
```

Because we need to iterate the extracted Tweets multiple times in the following analyses, we store the extracted Tweets in the *tweet_data* list to avoid unnecessary query operations on the MongoDB database. We also remove the URLs and new line characters in the Tweet texts with regular expressions.[58] Writing regular expressions is an intimidating work, even for seasoned programmers. Fortunately, we can use Amazon Q Developer or ChatGPT to help write the code.

```
tweet_data = []
url_pattern = r'http[s]?://
(?:[a-zA-Z]|[0-9]|[$-_@.&+]|[!*\\
(\\),]|(?:%[0-9a-fA-F][0-9a-fA-F]))+'
for tweet in result:
    text_without_urls = re.sub(url_pattern, '',
tweet['tweet']['text'])


tweet_data.append({'tweet_id':tweet['tweet']['id']
,'tweet_text':text_without_urls.replace('\n','')})
```

We can print out the number of Tweets in our tweet_data list:

```
print('Number of tweets: ',len(tweet_data))
```

With 434 Tweets in the *tweet_data* list, we are ready to set up the OpenAI APIs. First, we need to decide the *model type* and *temperature*:

- Token estimate: LLMs can process a certain number of words in each request, quantified as *tokens*. In English, 100 tokens are roughly 75 words, and you can get a more accurate estimate at *Tokenizer*.[59]
- Model type: The latest models generally outperform the old ones regarding token limitations and content accuracy, but they are more expensive. Please check the OpenAI documents for differences between the models. [60]
- Temperature: Lower temperatures produce more consistent outputs, while higher temperatures generate more diverse and creative results. The value can range from 0 to 2. Please see the OpenAI API reference for more details.[61]

Next, we can create a help function, *openai_help*, to pass the prompts to LLMs. This section uses gpt-4o,[62] which was released on May 13, 2024.

```
from openai import OpenAI
client = OpenAI(api_key=openai_api_key)
model="gpt-4o"
temperature=0
def openai_help(prompt, model=model, temperature
=temperature ):
    messages = [{"role": "user", "content": prompt}]
    response = client.chat.completions.create(
            model=model,
            messages=messages,
            temperature=temperature
    )
    return response.choices[0].message.content
```

We can now use LLMs to analyze the extracted Tweets by asking for specific tasks in prompts and save the results back to the MongoDB cluster. Below are some examples; please feel free to design different prompts.

- Sentiment analysis can help quickly summarize product reviews or customer feedback. For each Tweet in the tweet_data list, we ask the model to analyze the sentiment and specify the possible outputs, e.g., positive, neutral, or negative. We pass the prompts with the *openai_help* function to the LLM and store the responses in the MongoDB cluster. Specifically, we use an UPDATE operator to store the detected sentiment as a new field in each Tweet document.

```
for tweet in tqdm(tweet_data):
  prompt = f"""
  What is the sentiment of the following tweet,
  tweet text: {tweet['tweet_text']}
  return the result with one word as positive,
  neutral, or negative
  """
#   print(prompt)
  try:
        sentiment_result =openai_help(prompt)
#      print(sentiment_result)
        tweet_collection.update_one(
            {'tweet.id':tweet['tweet_id']},
            {"$set":{'tweet.sentiment':sentiment_res
            ult}}
        )
  except:
        pass
```

- Language translation allows us to understand texts in foreign languages. Most Tweets are written in English, with fewer in other languages. We ask LLM to translate the Tweet texts into Chinese in this analysis. We then store the results in the Tweet documents. Please feel free to translate the Tweets into another language.

```
for tweet in tqdm(tweet_data):
    prompt = f"""
    Translate the following tweet into Chinese
    tweet text: {tweet['tweet_text']}
    """
      # print(prompt)
    try:
        translate_result =openai_help(prompt)
          # print(translate_result)
        tweet_collection.update_one(
            {'tweet.id':tweet['tweet_id']},
            {"$set":{'tweet.translate':translate_res
            ult}}
    )
    except:
      pass
```

- Emotion identification can quickly flag reviews or social media posts that contain certain emotional tones and respond accordingly. We can ask LLM to identify Tweets expressing anger and store the result as a Boolean value in the Tweet documents. Please feel free to identify different emotions, such as happiness or depression.

```
for tweet in tqdm(tweet_data):
     prompt = f"""
     Does the following tweet express anger?
     Provide the result as either True or False.
     tweet text: {tweet['tweet_text']}
     """
       # print(prompt)
     try:
       emotion_result =openai_help(prompt)
         # print(emotion_result)
       tweet_collection.update_one(
           {'tweet.id':tweet['tweet_id']},
```

```
            {"$set":{'tweet.anger':emotion_result}}
        )
    except:
      pass
```

- Entity extraction allows analysts to filter documents and identify specific information quickly. In this example, we ask LLM to extract the names of people or organizations from each Tweet. We also instruct the model to organize the outputs in a JSON document and embed the document into the Tweet document.

```
for tweet in tqdm(tweet_data):
    prompt = f"""
    Identify persons or organizations from the
    following tweet,
    tweet text: {tweet['tweet_text']},
    format the response as a json object with person
    and organization as the keys
    Do not wrap the json codes in JSON markers
    If the information is not presented, use
    "unknown".
    """
        #  print(prompt)
    try:
        extract_result =openai_help(prompt)
        # print(extract_result)
        tweet_collection.update_one(
            {'tweet.id':tweet['tweet_id']},
{"$set":{'tweet.extracted_item':json.loads(extract
_result)}}
            )
    except:
        pass
```

- Text summarization improves analysts' efficiency by providing a short overview of large documents. We can specify the number of Tweets to be summarized in each batch based on the LLM token limit and instruct the model to focus on a specific perspective or limit the number of words in the summary. We then save analyzed Tweet IDs and summation into a separate MongoDB collection named *summary_result*.

```
# Define the batch size
batch_size = 500
start_index = 0
while start_index < len(tweet_data):
    batch = tweet_data[start_index:start_index +
    batch_size]
    tweet_id_list =[]
    tweet_text_summary =''
    for tweet in batch:
        tweet_id_list.append(tweet['tweet_id'])
        tweet_text_summary = tweet_text_summary+'
        .'+tweet['tweet_text']
    prompt = f"""
    Summarize the following tweets in at most 50
words,
    tweet text: {tweet_text_summary,}
    """
# print(prompt)
    try:
        summary_result =openai_help(prompt)
        tweet_summary = db.tweet_summary
        tweet_summary.insert_one({'id_list':tweet_id_
        list,
            'tweet_text_summary':summary_result})
        print(summary_result,'\n')
    except:
        pass
    start_index += batch_size
```

Below is the summary of the 434 Tweets:

```
The tweets discuss various election-related topics,
including election security concerns, court rulings
on mail-in ballots, disbarment of lawyers involved
in election disputes, accusations of election fraud,
and political commentary on Trump and Biden. Key
events include John Eastman's disbarment and a
significant court ruling in Pennsylvania.
```

After completing the Twitter data analysis, close the database connection in Python, upload the notebook to the GitHub repository, and stop the JupyterLab instance.

We can now view the results in the MongoDB *demo* database and create a dashboard with Atlas Charts. We need to log in to MongoDB Atlas,[63] open the *Database* page, and *Browse Collections*. In the *demo* database, we can see both the *tweet_collection* and *tweet_summary*. To create a dashboard, click the *Charts* button at the top and *add a dashboard*. We can provide a name and description for the dashboard and *add charts*.

Before creating a visualization, we need to choose the data source. We can select the *tweet_collection* in the *demo* database, as shown in Figure 6.20. Unlike QuickSight, which supports different data sources, Atlas Chart can only visualize MongoDB data in the same project.

Figure 6.21 shows the Classic Atlas Chart editing interface. On top of the editor, we can switch data sources or filter the data with document queries or aggregation pipelines. Like QuickSight, we can drag fields from the left side, select chart types, edit aggregation functions and filters, and customize chart properties in the middle. The created chart is displayed on the right side. Three suggested charts are already created automatically.

First, we want to see the number of Tweets with different *sentiments*. We can use a *Stacked Bar* chart and drag the *Tweet ID* field to the *X-axis* with the COUNT function. We also drag the *sentiment* field to the *Y-axis* and bring the *anger* field to the *Series*. Because we only analyzed part of the datasets, many Tweet documents don't have the sentiment and anger fields. Therefore, we need to add *filters* to exclude the Tweets that don't have those fields. As shown in Figure 6.22, most analyzed Tweets are classified as negative by LLM, and only a small number of Tweets are positive. All the anger Tweets are classified as negative.

We can *Save and close* this chart and create a new chart using the same data source. The second chart displays the number of Tweets created at different times. Because the *created_at field* is a string, we need to convert it to *date* by clicking the three dots next to the field name and choosing *Convert type*. We then select the *Discrete Line* chart as the chart type and bring the *created_at* to the X-axis and the *Tweet ID* to the Y-axis. We select *Minute* as the bin range on the X-axis and select COUNT as the aggregation function on the Y-axis. We can *Customize* the created_at field format and convert the time zone from

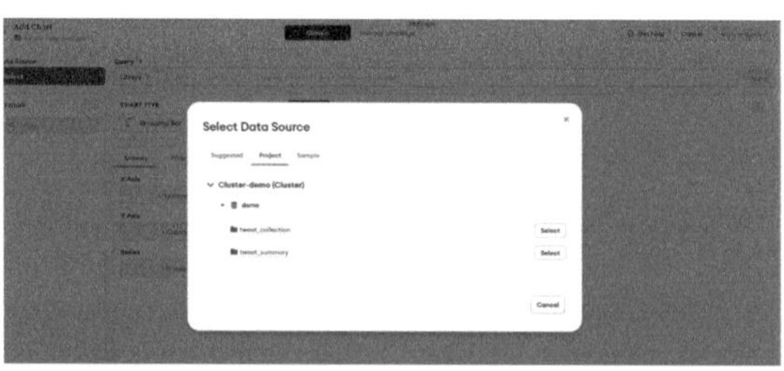


**FIGURE 6.20**
Select Data Source on Atlas Chart

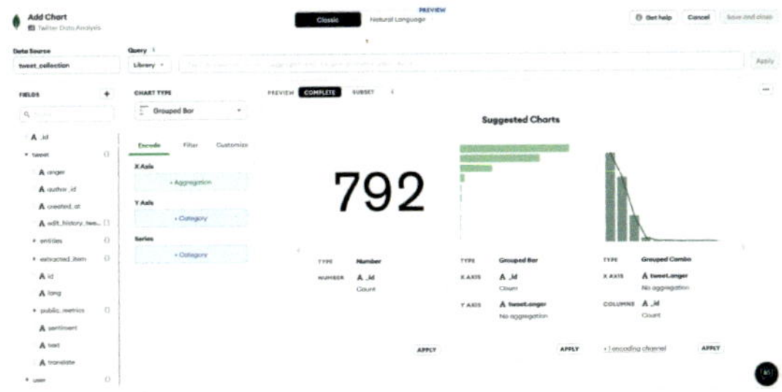


**FIGURE 6.21**
Classic Interface of Atlas Chart

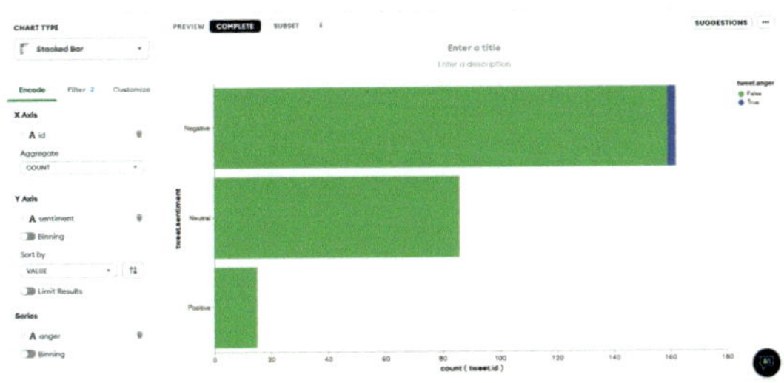


**FIGURE 6.22**
Tweet Sentiment and Emotion

Coordinated Universal Time (UTC) to Eastern Time. Figure 6.23 shows that we collected over 700 Tweets posted in less than 10 minutes.

Next, we want to see who the extracted persons are compared to the extracted organizations. We can choose the *Heatmap* as the chart type and drag the *extracted_item.person* field to the X-axis and the *extracted_item.organization* field to the Y-axis. We also need to unwind both fields as they contain lists and limit the results to show the top 10 people. We then drag the Tweet *ID* to the *Intensity* with a COUNT aggregation function. Finally, we filter out the null values in the person and organization fields to exclude the Tweets that have not been analyzed. Figure 6.24 illustrates the frequently mentioned persons and organizations related to the 2024 US presidential election. For example, *Trump* is mentioned together with *Fox News*, and Biden is commonly associated with *AP*, which may be Associated Press News.

We also want to display user locations on a map. We can create a *Choropleth* map using the *user.location* field as the map location, and the COUNT of the *user.id* as the map color. In the map customization, we choose a *diverging color* scheme and select *US States* as the shape scheme. Figure 6.25 displays the created map where states with high values are assigned a dark blue color, and states with low values are assigned a dark red color. Please note: The choropleth map is unsuitable for displaying quantities, such as the number

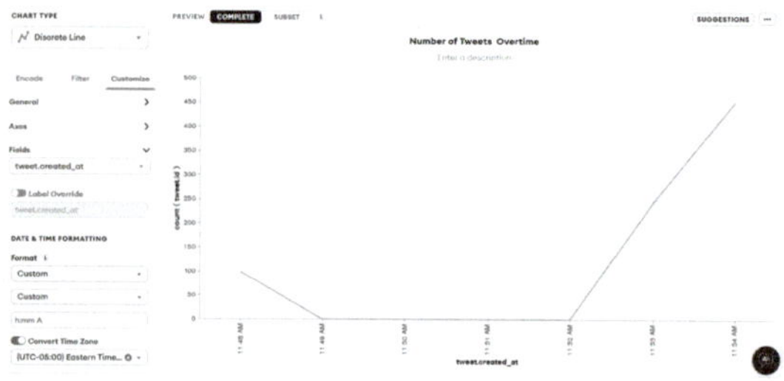


**FIGURE 6.23**
Number of Tweets Over Time

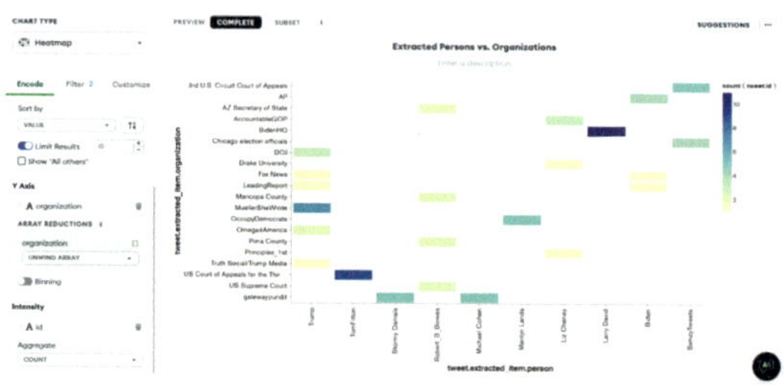


**FIGURE 6.24**
Extracted Persons vs. Organizations

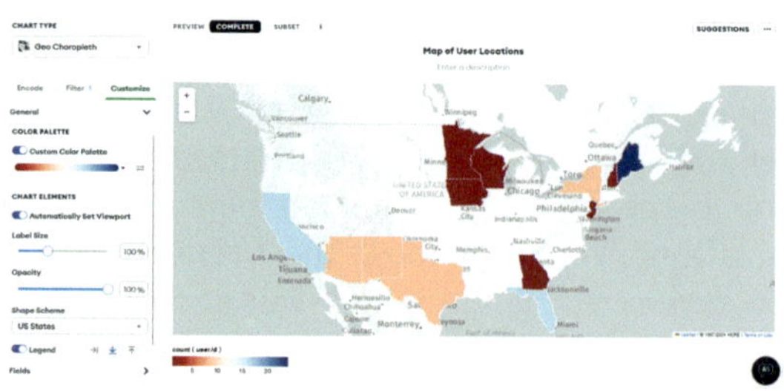


**FIGURE 6.25**
Map of User Locations

of users. However, the proportional symbol map is incompatible with the collected Tweet data, so we use a choropleth map here.

We now switch to the Natural Language editor and write a prompt to visualize the data.

```
Find the top 10 users with the most followers on a
bar chart
```

After clicking the *Generate chart* button multiple times, we received a bar chart showing the top 10 *usernames* sorted by the *followers_count*. As shown in Figure 6.26, *aajtak* is the most influential Twitter user, with more than 20 million followers. The GenAI model also adds a chart title called *Social Titans*.

When all the charts are completed, we can add interactivity to the dashboard by defining a *dashboard filter*[64] and then enabling individual charts as filters.[65] Figure 6.27 shows the dashboard, where the neutral Tweets are selected on the top bar chart and serve as filters on the other charts. The dashboard can be shared publicly or privately with selected users. We can set the time interval when the dashboard refreshes or emails are sent to report new collected or analyzed Tweets.

The three case studies show that GenAI models provide decent Python coding in data analysis but limited help with data visualizations. Specifically, OpenAI models help write accurate Python codes and SQL statements but generate inconsistent output when analyzing Tweets. For example, the identified sentiments can be *negative* or *Negative*. *Tom Fitton* and *TomFitton* are considered different persons extracted from the Tweets. We can improve the model performance by prompt engineering or cleaning the outputs before data visualizations.

Amazon Q and MongoDB Natural Language Charts in BI tools create more errors than applicable visualizations. Many times, even the system-suggested prompts can't work correctly. Given that those functions were

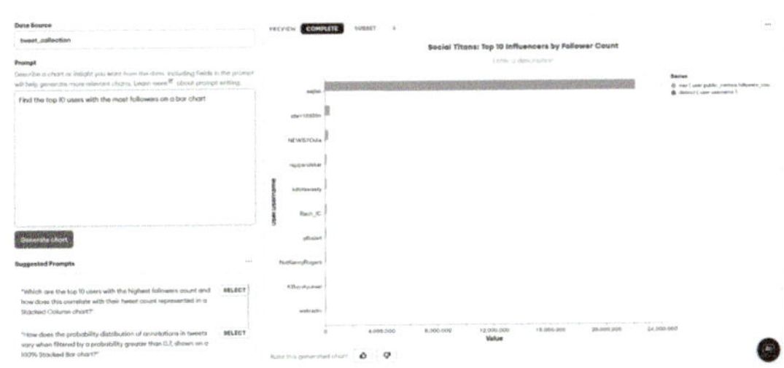

**FIGURE 6.26**
Top 10 Influencers by Follower Count

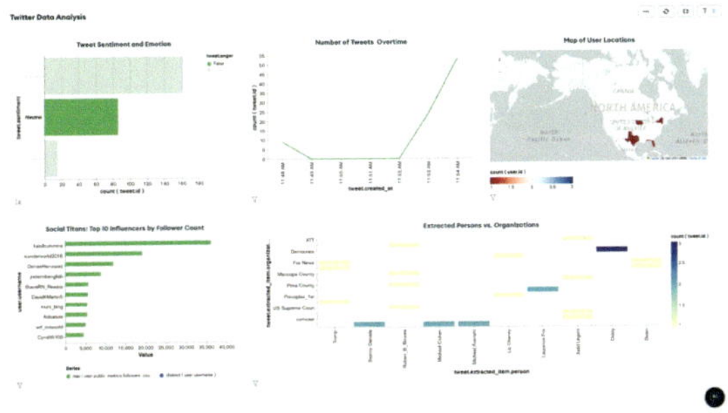

**FIGURE 6.27**
Interactive MongoDB Dashboard

recently released and in preview status, we expect their performance to improve soon.

Although GenAI is a robust data analysis and visualization assistant, it does access our data.[66] When analyzing data with AI, we need to follow institutional data regulations and privacy policies to protect data privacy. We must also comply with domestic or international law, such as the US Health Insurance Portability and Accountability Act (HIPPA),[67] the Family Educational Rights and Privacy Act (FERPA),[68] or The EU Artificial Intelligence Act[69] when dealing with sensitive information.

## Notes

1. https://help.tableau.com/current/blueprint/en-us/bp_why_visual_analytics.htm
2. https://aws.amazon.com/quicksight/
3. https://www.tableau.com/
4. https://www.microsoft.com/en-us/power-platform/products/power-bi
5. https://www.esri.com/en-us/arcgis/products/arcgis-pro/overview
6. https://www.tableau.com/viz-gallery
7. https://open.maricopa.edu/gist/chapter/3-2-symbology/
8. https://www.sciencedirect.com/topics/physics-and-astronomy/moire-effect
9. https://pro.arcgis.com/en/pro-app/latest/help/mapping/layer-properties/proportional-symbology.htm
10. https://www.theguardian.com/news/datablog/2013/jul/24/why-you-should-never-trust-a-data-visualisation
11. https://www.colourblindawareness.org/colour-blindness/
12. https://www.tableau.com/blog/examining-data-viz-rules-dont-use-red-green-together
13. https://www.dataplusscience.com/files/Shaffer%204C%20-%20Clean%20Examples.pdf.
14. https://www.storytellingwithdata.com/blog/2012/09/some-finer-points-of-data-visualization
15. https://www.tableau.com/learn/whitepapers/which-chart-or-graph-is-right-for-you
16. https://www.nngroup.com/articles/text-scanning-patterns-eyetracking/
17. https://www.utsa.edu/marcomstudio/resources/style-guides/writing-guide/data-storytelling.html
18. https://www.actionrocket.co/blog/10-tips-when-combining-fonts
19. https://help.tableau.com/current/blueprint/en-us/bp_visual_best_practices.htm
20. https://aws.amazon.com/quicksight
21. https://www.mongodb.com/products/platform/atlas-charts
22. https://github.com/jupyterlab/jupyter-ai
23. https://aws.amazon.com/q/

24. https://www.mongodb.com/docs/charts/chart-type-reference/natural-language-charts/
25. https://pypi.org/project/psycopg2/
26. https://pypi.org/project/pymongo/
27. https://pandas.pydata.org/
28. https://matplotlib.org/
29. https://docs.aws.amazon.com/wellarchitected/latest/analytics-lens/design-principle-15.html
30. https://aws.amazon.com/q/developer
31. https://openai.com/chatgpt/
32. https://colab.research.google.com/
33. https://huggingface.co/
34. https://gemini.google.com/
35. https://openai.com/index/gpt-4/
36. https://openai.com/
37. https://pypi.org/project/jupyter-ai/
38. https://pypi.org/project/langchain-openai/
39. https://github.com/xbwei/Social-Data-Analytics-in-the-Cloud-with-AI
40. https://jupyter-ai.readthedocs.io/en/latest/users/index.html
41. https://github.com/jupyterlab/jupyter-ai/tree/main/examples
42. https://jupyter-ai.readthedocs.io/en/latest/users/index.html#model-providers
43. https://pandas.pydata.org/docs/reference/api/pandas.DataFrame.pct_change.html
44. https://openai.com/index/gpt-4/
45. https://openai.com/index/gpt-3-apps/
46. https://aws.amazon.com/blogs/business-intelligence/amazon-q-is-now-generally-available-in-amazon-quicksight-bringing-generative-bi-capabilities-to-the-entire-organization/
47. https://aws.amazon.com/q/
48. https://docs.aws.amazon.com/quicksight/latest/user/spice.html
49. https://aws.amazon.com/quicksight/pricing
50. https://www.census.gov/library/stories/2023/12/happy-new-year-2024.html
51. https://gemini.google.com
52. https://www.mongodb.com/products/platform/atlas-charts
53. https://www.mongodb.com/docs/charts/chart-type-reference/natural-language-charts
54. https://www.mongodb.com/docs/charts/release-notes/
55. https://pypi.org/project/pymongo/
56. https://github.com/openai/openai-python
57. https://github.com/xbwei/Social-Data-Analytics-in-the-Cloud-with-AI
58. https://docs.python.org/3/library/re.html
59. https://platform.openai.com/tokenizer
60. https://platform.openai.com/docs/models
61. https://platform.openai.com/docs/api-reference/chat/create#chat-create-temperature
62. https://openai.com/index/hello-gpt-4o/
63. https://account.mongodb.com/account/login

64. https://www.mongodb.com/docs/charts/dashboards/view-dashboard-data/dashboard-filtering/
65. https://www.mongodb.com/docs/charts/dashboards/view-dashboard-data/interact-with-charts/
66. https://www.mongodb.com/docs/charts/ai-and-data-usage-information/
67. https://www.hhs.gov/hipaa/index.html
68. https://www2.ed.gov/policy/gen/guid/fpco/ferpa/index.html
69. https://artificialintelligenceact.eu/

## References

Anscombe, F. J. 1973. "Graphs in Statistical Analysis." *The American Statistician* 27 (1): 17–21. https://doi.org/10.1080/00031305.1973.10478966.

Brown, Richard O., and Donald I. A. MacLeod. 1997. "Color Appearance Depends on the Variance of Surround Colors." *Current Biology* 7 (11): 844–49. https://doi.org/10.1016/S0960-9822(06)00372-1.

Chris Pidsley. 2022. "10 Tips When Combining Fonts." Action Rocket. September 2022. https://www.actionrocket.co/blog/10-tips-when-combining-fonts.

Dent, Borden, Jeff Torguson, and Thomas Hodler. 2008. *Cartography: Thematic Map Design*. 6th edition. Boston: McGraw-Hill Education.

Hockley, William E. 2008. "The Picture Superiority Effect in Associative Recognition. *Memory & Cognition* 36 (7): 1351–59. https://doi.org/10.3758/MC.36.7.1351.

Kirk, Andy. 2012. *Data Visualization: A Successful Design Process: A Structured Design Approach to Equip You with the Knowledge of How to Successfully Accomplish Any Data Visualization Challenge Efficiently and Effectively*. Birmingham: Packt Pub Ltd.

Project Jupyter. n.d. "Jupyter AI—Jupyter AI Documentation." Accessed July 12, 2024. https://jupyter-ai.readthedocs.io/en/latest/.

Salesforce, Inc. n.d. "Visual Best Practices." Accessed July 11, 2024. https://help.tableau.com/current/blueprint/en-us/bp_visual_best_practices.htm.

Shaikh, A. D., B. Chaparro, and D. Fox. 2006. "Perception of Fonts: Perceived Personality Traits and Uses." https://www.semanticscholar.org/paper/Perception-of-fonts%3A-Perceived-personality-traits-Shaikh-Chaparro/9769026987797773c2633c61c202560ffbb24819.

Steele, Julia, and Noah Iliinsky. 2010. *Beautiful Visualization*. 1st edition. Beijing Köln: O'Reilly Media.

Tracy Rodgers. n.d. "Which Chart or Graph Is Right for You? A Guide to Data Visualization." Accessed July 11, 2024. https://www.tableau.com/learn/whitepapers/which-chart-or-graph-is-right-for-you.

Tufte, Edward R. 2001. "Graphical Excellence." In *The Visual Display of Quantitative Information*, edited by Edward R. Tufte, 2nd edition. Cheshire, CT: Graphics Press.

# 7

# *Conclusion*

## Responsible Data Analytics

Now, we have completed a data analytics journey in the cloud with AI: We started by collecting data from various sources using APIs and stored it in different data containers, such as relational or NoSQL databases. We then queried the collected data using SQL or MongoDB CRUD language and visualized it using open-source Python libraries or business intelligence tools. Readers can access the demo codes on our GitHub repository.[1] During the journey, Generative AI powered by foundation models makes each step faster and easier. We should collect, analyze, and visualize data *responsibly* with the assistance of Generative AI.

*Public data are not equal to public use*: Even if a dataset can be technically obtained, it is essential to always check and follow the data resource guidelines before using or collecting it. This includes verifying permissions and adhering to ethical and legal standards. For instance, before crawling data from any website, we should examine the robots.txt file to understand the site's restrictions on automated access. Similarly, when applying for access to APIs, it is crucial to thoroughly read and understand the developer terms and policies. These steps ensure compliance with the data provider's rules and help maintain the integrity and legality of our data collection practices. For example, we can't redistribute the collected Tweets per the X Developer Policy.[2] As data becomes more valuable in the era of Generative AI, violating data usage policies is not only an ethical issue but a legal issue.

"Security in the Cloud": The cloud is never secure. For example, cyberattacks on health systems cost money and lives (McGlave, Neprash, and Nikpay, 2023). While cloud providers are responsible for maintaining the physical infrastructures of cloud services, users are tasked with the protection and management of their cloud resources. This includes configuring security settings, managing access controls, and ensuring data privacy. Users must also monitor and optimize the performance of their cloud-based applications, regularly update software, and implement backup and recovery solutions to safeguard their data. By effectively managing these responsibilities, users can leverage the full potential of cloud services while maintaining robust

DOI: 10.1201/9781003437611-7

security and operational efficiency. In the AWS *Shared Responsibility Model*,[3] AWS is responsible for *the security of the cloud*. Customers are also responsible for *the security in the cloud*, and more flexibility in the cloud comes with higher responsibility. As best practices, we should encrypt sensitive data, store credentials with secret manager services, rotate tokens regularly, and enable multi-factor authentication (MFA)[4] when possible.

*We are responsible for analysis results*: All datasets inevitably contain errors, and all visualizations inherently distort data to some extent. These errors can stem from various sources, including measurement inaccuracies, data entry mistakes, or missing information. Additionally, visualizations, while useful for interpreting data, often introduce distortions through the choice of scales, colors, and graphical representations. Moreover, data itself may carry biases, which can arise from the way it is collected, processed, or interpreted. These biases can be due to sampling methods, cultural influences, or preexisting assumptions embedded in the data collection process (Ye et al., 2023). Although Generative AI can create texts or images, the analyst is still responsible for processing and cleaning data and creating visualizations that reflect true stories (GitHub, Inc., n.d.; OpenAI, n.d.). For example, the OpenAI Usage Policy prohibits creating or disseminating misinformation or misleading content.[5] GitHub claims that Copilot is a tool that can't replace human programming.[6] We should realize that Generative AI still does not understand the content it generates but instead predicts it based on extensive training datasets, which often contain biases and errors. Therefore, we should thoroughly examine AI-generated content and rigorously test the codes or results, even if they appear valid.

*We need to balance economic benefits with environmental impact*: While cloud computing provides numerous benefits, it also introduces challenges that must be carefully managed. From an economic perspective, cloud technology can lower costs by eliminating the need for significant initial investments in hardware and maintenance. It allows organizations to scale resources based on demand, thereby optimizing spending (Purcell, 2014). However, reliance on cloud service providers entails ongoing operational expenses and potential cost variability. For example, a CrowdStrike security update triggered a global Microsoft outage on July 19, 2024, causing flight cancellations, first responder disruptions, and business paralysis.[7] Additionally, centralized data storage can necessitate increased spending on cybersecurity to protect sensitive information (Isakov et al., 2024). Environmentally, cloud technology has a substantial impact. Data centers, which power cloud services, consume large quantities of electricity and water, contributing to carbon emissions and environmental degradation (Mark, John, Bommu, 2024). Despite improvements in energy efficiency and the adoption of renewable energy sources, the overall environmental footprint remains a concern. The entire lifecycle of data centers, from construction to decommissioning, involves significant resource use and waste generation.

## Big Data with Large Models

Big data created by human beings powers the training of foundational models, which are sophisticated AI systems designed to understand and generate human-like text and other outputs. These foundational models, in turn, facilitate the analysis and consumption of this human-generated big data, making it easier to extract insights, identify patterns, and make informed decisions across various fields. We may eventually reach a point where AI outperforms human capabilities, namely artificial general intelligence (AGI). According to a Bloomberg report[8] on July 11, 2024, OpenAI classified the capabilities of artificial intelligence into five levels: conventional AI, reasoning AI, autonomous AI, innovation AI, and organizational AI. While OpenAI is currently at the first level, the AI technology may evolve to level five to "do the work of an organization" or even reach the level of AGI.

While it is hard to project when AGI will arrive, we believe that *data analytics is becoming increasingly accessible*: Coding or data querying should not be a constraint in data analytics, and analysts can spend more time asking questions and exploring new ideas. Open-source foundation models, such as Llama,[9] have become available, and academia and non-profit communities can collaborate on models, datasets, and applications on platforms like Hugging Face.[10]

We should *Keep Learning* as new models and technologies are constantly invented (Amazon Web Services, Inc. 2023). For example, we can add *watermarking*[11] to AI-generated images to reduce the spreading of disinformation. AI agents[12] can independently collect and analyze data and make predictions based on human-set goals. We can also analyze data in real-time analytics platforms[13] or immersive environments with virtual reality (VR) devices.[14]

As mentioned in the first chapter, we recommend the AWS Academy[15] courses on learning cloud computing and DeepLearning.AI[16] to explore Generative AI technologies. We will keep updating the latest data analytics techniques on our website, LBSocial.net.[17]

## Notes

1. https://github.com/xbwei/Social-Data-Analytics-in-the-Cloud-with-AI
2. https://developer.x.com/en/developer-terms/policy
3. https://aws.amazon.com/compliance/shared-responsibility-model/
4. https://aws.amazon.com/what-is/mfa/
5. https://openai.com/policies/usage-policies/
6. https://docs.github.com/en/copilot/responsible-use-of-github-copilot-features
7. https://www.wsj.com/tech/cybersecurity/crowdstrike-outage-software-patch-78d05df2

8. https://www.bloomberg.com/news/articles/2024-07-11/openai-sets-levels-to-track-progress-toward-superintelligent-ai
9. https://llama.meta.com/
10. https://huggingface.co/
11. https://aws.amazon.com/blogs/machine-learning/announcing-new-tools-and-capabilities-to-enable-responsible-ai-innovation/
12. https://aws.amazon.com/what-is/ai-agents/
13. https://www.mongodb.com/resources/basics/real-time-analytics-examples
14. https://www.apple.com/apple-vision-pro/
15. https://aws.amazon.com/training/awsacademy/
16. https://www.deeplearning.ai/
17. https://www.lbsocial.net/

## References

Amazon Web Services, Inc. 2023. "Announcing New Tools and Capabilities to Enable Responsible AI Innovation | AWS Machine Learning Blog." November 29, 2023. https://aws.amazon.com/blogs/machine-learning/announcing-new-tools-and-capabilities-to-enable-responsible-ai-innovation/.

GitHub, Inc. n.d. "Responsible Use of GitHub Copilot Features." GitHub Docs. Accessed July 11, 2024. https://docs.github.com/en/copilot/responsible-use-of-github-copilot-features.

Isakov, Abror, Fakhridin Urozov, Shahboz Abduzhapporov, and Mukhlisa Isokova. 2024. "Enhancing Cybersecurity: Protecting Data in the Digital Age." *Innovations in Science and Technologies* 1 (1): 40–49.

Mark, John, and Revathi Bommu. 2024. "Tackling Environmental Concerns: Mitigating the Carbon Footprint of Data Transmission in Cloud Computing." *Unique Endeavor in Business & Social Sciences* 3 (1): 99–112.

McGlave, Claire C., Hannah Neprash, and Sayeh Nikpay. 2023. *Hacked to Pieces? The Effects of Ransomware Attacks on Hospitals and Patients.* SSRN Scholarly Paper. Rochester, NY. https://doi.org/10.2139/ssrn.4579292.

OpenAI. n.d. "Usage Policies." Accessed July 11, 2024. https://openai.com/policies/usage-policies/.

Purcell, Bernice M. 2014. "Big Data Using Cloud Computing." *Journal of Technology Research* 5: 1.

Ye, Xinyue, Galen Newman, Chanam Lee, Shannon Van Zandt, and Dawn Jourdan. 2023. "Toward Urban Artificial Intelligence for Developing Justice-Oriented Smart Cities." *Journal of Planning Education and Research* 43 (1): 6–7.

# Index

For Product Safety Concerns and Information please contact our EU representative GPSR@taylorandfrancis.com Taylor & Francis Verlag GmbH, Kaufingerstraße 24, 80331 München, Germany

to undergo odontoblastic differentiation, support the formation of new vascular networks, and contribute to the re-establishment of neural elements, ensuring both structural integrity and functional recovery of the regenerated tissue. Dental-derived stem cell populations, including DPSCs, SCAP, and PDLSCs, are central to these phenomena. Their capacity to differentiate into lineage-appropriate cell types, secrete trophic factors, and respond dynamically to the surrounding microenvironment enables them to regulate inflammation, promote tissue remodeling, and guide the reconstitution of vital pulp tissue.

## 5.4 CELLULAR PROCESSES INVOLVED IN PULP AND ROOT REGENERATION

Odontoblastic differentiation represents the initial and most fundamental step in the regeneration of pulp and root structures. During this phase, stem cells adopt an odontoblast-like phenotype capable of depositing new dentin. The process is initiated by a series of molecular signals that include BMPs, TGF-β, and components of the Wnt signaling pathway. Exposure to dentin-derived molecules such as dentin sialophosphoprotein and dentin matrix protein 1 provides additional inductive signals that reinforce odontogenic commitment. These factors converge on key transcriptional regulators, notably Runx2 and Osterix, which direct the expressions of genes associated with odontoblast differentiation.

Following commitment, newly formed odontoblast-like cells position themselves along the periphery of the pulp space, where they begin synthesizing a dentin matrix rich in type I collagen and hydroxyapatite. Subsequent mineralization of this matrix results in the formation of reparative or reactionary dentin, contributing to the restoration of structural continuity in teeth affected by injury or disease. The capacity of dental stem cells to transition into odontoblast-like cells and support organized dentinogenesis is central to the success of regenerative endodontic procedures, providing both anatomical restoration and re-establishment of biological function within the pulp–dentin complex.

## 5.5 MOLECULAR MECHANISMS REGULATING ODONTOGENESIS, VASCULARIZATION, AND INNERVATION

The re-establishment of vascular and neural networks within regenerated pulp tissue is essential for maintaining its long-term vitality and biological function. Angiogenesis is initiated primarily through the secretion of pro-angiogenic mediators such as VEGF and platelet-derived growth factor by dental-derived stem cells. These signaling molecules recruit and stimulate endothelial cells, promoting their migration, proliferation, and organization into new vascular structures. The resulting blood vessels restore oxygen and nutrient supply to the regenerating tissue, which is critical for the survival and activity of newly differentiated odontoblast-like cells. Innervation plays an equally essential role in restoring the physiological function of regenerated pulp tissue as sensory perception and protective reflexes rely on its connection to the peripheral nervous system. Dental stem cells contribute to this process by secreting neurotrophic factors such as nerve growth factor and brain-derived neurotrophic factor, which support the survival, maturation, and directed migration of neural progenitor cells. These signaling molecules facilitate the extension of axonal processes into the regenerating pulp, enabling the establishment of functional neural networks. The presence of active nerve fibers restores nociceptive and mechanosensory ability and enhances local immune surveillance, thereby improving the resistance of the regenerated tissue to microbial insult.

The coordination of odontoblastic differentiation, angiogenesis, and innervation is governed by the microenvironment within the root canal system. The ECM plays a central role by providing structural organization and delivering biochemical signals that regulate stem cell behavior. Biomaterial scaffolds used in regenerative procedures amplify these effects by mimicking the ECM architecture and offering a 3D template that supports cell attachment, proliferation, and lineage commitment. When combined with appropriate growth factors and other bioactive molecules, these scaffolds significantly enhance the efficiency and predictability of stem cell-mediated pulp–dentin regeneration.

## 5.6 ROLE OF GROWTH FACTORS AND SIGNALING MOLECULES

The regenerative potential of stem cell-based therapies in endodontics is governed by a coordinated network of growth factors and signaling molecules that regulate the key cellular events underlying pulp–dentin complex repair. These biochemical mediators influence stem cell proliferation, migration, differentiation, and ECM production, thereby shaping a microenvironment that supports effective tissue restoration. Among the numerous signaling pathways involved, BMPs, TGF-β, and VEGFs are among the most important regulators, each contributing distinct yet complementary roles in the regeneration of dental tissues.

BMPs, particularly BMP-2 and BMP-7, are central to the induction of odontoblastic differentiation. As members of the TGF-β superfamily, BMPs initiate intracellular signaling through Smad-dependent and MAPK pathways upon receptor binding. These cascades activate transcriptional